LANCHESTER LIBRARY

3 8001 00586 1061

WITHDRAWN

LANCHESTER LIBRARY, Coventry University
Gosford Street, Coventry CVI 5DD Telephone 024 7688 7555

_ 4 JUL 2012

This book is due to be returned not later than the date and
time stamped above. Fines are charged on overdue books

LIFETI

Chris Go

Habint

Habinteg Housing Association and the publisher make every effort to ensure the accuracy and quality of information and guidance when it is first published. However, we can take no responsibility for the subsequent use of this information, nor for any errors or omissions that it may contain.

IHS BRE Press supplies a wide range of building and construction-related information products from BRE and other respected organisations. Details are available from:

www.brebookshop.com
or
IHS BRE Press
Willoughby Road
Bracknell RG12 8FB
Tel: 01344 328038
Fax: 01344 328005
Email: brepress@ihs.com

Published by IHS BRE Press for
Habinteg Housing Association
Holyer House
20–21 Red Lion Court
London EC4 3EB
www.habinteg.org.uk

Requests to copy any part of this publication should be made to the publisher:
IHS BRE Press
Garston, Watford WD25 9XX
Tel: 01923 664761
Email: brepress@ihs.com

Printed on paper sourced from responsibly managed forests

The publisher accepts no responsibility for the persistence or accuracy of URLs referred to in this publication, and does not guarantee that any content on such websites is, or will remain, accurate or appropriate.

Front cover photos:
    Left, Consort Road Housing, London
    © copyright Edmund Sumner/VIEW
    Top right, Anne Mews, Barking
    © copyright Tim Soar and Allford Hall
    Monaghan Morris
    Bottom right, Parc Ecologic, Redruth.
    © copyright Cloud Nine
Back cover photo:
    Arundel Square, London
    © copyright Tim Crocker and
    Pollard Thomas Edwards architects

Index compiled by Paul Nash

Coventry University Library

EP 100
© Copyright Habinteg Housing
Association 2011
First published 2011
ISBN 978-1-84806-188-0

FOREWORD

Designing the built environment, and people's homes within that environment, needs to be right first time if we are to achieve high livability and sustainability standards. People cannot take their home back to the shop because it 'doesn't quite fit', or because it 'doesn't quite suit me'. Few of us will ever have the luxury of commissioning bespoke homes, tailored to our specific needs. Most will have a limited choice, restrained by a budget, and will therefore be buying, or renting, from an 'off the peg' range. It therefore makes sense for the design of all housing to consider, as far as practicable, the diverse needs of the widest range of people who may take up residence — either initially, or during the life of the home.

In addition, housing choices are usually long term. Moving home, or altering the home, can be an expensive, disruptive, and stressful experience. Sometimes a 'forced move' could be prevented if the dwelling had a degree of flexibility designed in that enabled simple adjustments and adaptations to be made to cater for a change in the household's needs. As a household ages, these needs may often relate to reduced mobility or agility, and planning for this in the original design, while initially seeming irrelevant, may prove a huge bonus to the household later on. At the least, it will offer the household choices that would not otherwise exist. For example, providing a slightly wider space to get in and out of the car (essential for those needing to use mobility aids) is also helpful to parents lifting infants in and out of the car. Providing wider parking spaces for 'parents and toddlers' is now standard at the supermarket so it makes sense to have the same convenience when arriving home.

As well as the obvious benefits to the potential occupiers and their visitors, increasing and mainstreaming housing stock that enables simple adjustments to cater for people's changing needs offers cost benefits. These benefits could save the household the expense of a move, or the disruption and cost of a significant adaptation, or, looking at the broader picture, save significant public money if the flexibility designed in at the outset prevents the household needing specialist accommodation, extensions or significant adaptations.

This design guide discusses the Lifetime Homes Standard, a Standard that embodies the inclusivity, adaptability and value principles alluded to above. The Standard has become the industry's forerunner in the provision of housing that offers accessibility and flexibility in occupation. The guide explains the principles underlying the Standard, gives an overview of its history and a brief explanation of where it sits in current policy and regulation. Most importantly, in its seven distinct functional sections, it offers clear detailed explanation, and instruction, to those seeking to implement the Standard in designing and building new homes.

Wayne Hemingway
Designer and Chair of Building for Life
September 2011

HISTORY OF THE LIFETIME HOMES STANDARD

The originating idea for the Lifetime Homes Standard came from discussions in the late 1980s between the Helen Hamlyn Foundation and Habinteg Housing Association.

The term 'Lifetime Homes' was first adopted in connection with a project launched by the Helen Hamlyn Foundation in 1989, and applied at that time to the needs of older people.

Habinteg also had concerns about how inaccessible and inconvenient many new homes were for large sectors of the population and recognised that the concept initiated by the Helen Hamlyn Foundation could have a wider application that would fit neatly with its objectives of creating integrated and inclusive housing for a wide range of people with differing needs.

Discussions between the Helen Hamlyn Foundation and Habinteg broadened to others, including the Joseph Rowntree Foundation (JRF). The JRF brought together a number of housing experts in the early 1990s to form the JRF Lifetime Homes Group. This Group instructed a technical sub-group, consisting of architects, developers, access consultants and other industry professionals and practitioners, to develop a set of design standards for general needs housing, which would improve access and adaptability for a wide range of households with differing needs. The standards that evolved from this work became the 16 Lifetime Homes design criteria.

In 2008, the JRF transferred the lead role in the promotion, guidance and administration of the Standard to Habinteg. In the following year, Habinteg convened a new technical advisory group, comprising a range of practitioners and planners involved in housing design, development, access consultancy and the provision of adaptations. This group considered the 16 design criteria in great detail, with the aim of making it more practicable for volume developers to implement the Standard across the full range of modern house types and in both high- and low-density developments. At the same time, a research project was carried out on behalf of the UK Government, which canvassed the opinion of key private sector housing developers and architects with regard to implementing the Standard and the clarity of the guidance.

The findings from the research, although not published by the UK Government, were fed into the discussions of the technical advisory group, which then proposed a number of revisions to the 16 design criteria. The proposed revisions were released for public consultation and revisions to the original Lifetime Homes Standard were issued in 2010. The technical guidance in this Guide relates to this 2010 Standard.

In 2010, Habinteg was also officially granted the trademark for Lifetime Homes.

CONTENTS

ACKNOWLEDGEMENTS

Habinteg wishes to thank members of the Lifetime Homes Technical Advisory Group for their advice, knowledge and experience applied to the review of the Lifetime Homes Standard in 2010. The members of the Group were:

- Roger Battersby, PRP Architects
- David Bonnett, David Bonnett Associates
- Mike Donnelly, Habinteg
- Julie Fleck, Greater London Authority
- Chris Goodman, Habinteg
- Justin Halewood, BRE (Observer)
- Paul Jenkins, Taylor Wimpey
- Rachael Marshall, David Bonnett Associates
- Graham Nickson, Town and Country Planning Association
- Julia Park, Levitt Bernstein
- David Petherick, formerly of Department of Communities and Local Government
- Kate Sheehan, College of Occupational Therapists
- Andrew Shipley, Habinteg
- Rachel Smalley, Access Association
- Darryl Smith, Access Association
- Adam Thomas, Design Matters

1 INTRODUCTION

WHAT IS THE LIFETIME HOMES STANDARD?

The Lifetime Homes Standard[1] was established in the 1990s to incorporate a set of principles that are implicit in good contemporary housing design. Good housing design, in this context, is thoughtful, forward-looking design that maximises utility, independence and quality of life, while not compromising other design issues such as aesthetics or cost effectiveness.

The Standard is an expression of 'inclusive design'. It seeks to provide design solutions in general-needs housing that can meet the changing needs of the widest range of households. Some of these solutions should be included in the design from the outset, while others can be incorporated at a later stage through simple and cost-effective adaptation. This will give many households more choice over where they decide to live, the type of property they live in and the range of visitors they can readily accommodate.

Lifetime Homes properties are more convenient for most occupants and visitors, including those with less agility and mobility. The original design may negate the need for substantial alterations in order to make the dwelling suitable for a household's particular needs. The added convenience provided by the Standard is helpful to everyone in ordinary daily life, for example when carrying large and bulky items, or pushing child buggies.

The Standard is based on five overarching principles, as outlined below. These inform the rationale for the Lifetime Homes design requirements.

OVERARCHING PRINCIPLES

Principle 1: Inclusivity

An inclusive environment aims to assist use by everyone, regardless of age, gender or disability. It does not attempt to meet every need, but by considering the varying needs of individuals and households it aims to break down unnecessary barriers.

The design of a Lifetime Home removes the barriers to accessibility often present in other dwellings. The flexibility and adaptability within the design and structure enables a Lifetime Home to meet a diverse range of needs over time. A new development of Lifetime Homes therefore has the potential to provide for the widest cross-section of individuals within the general population. The high level of accessibility also offers greater 'visitability', for example to wheelchair users, so that people are less likely to be prevented from visiting due to the dwelling's design.

Principle 2: Accessibility

Inclusive design aims to give the widest range of people, including those with physical and/or sensory impairments, older people and children, convenient and independent access within the built environment (externally and internally) and also equal access to services.

A Lifetime Home is designed with particular attention to:
- the ease of approaching the home
- circulation within the home
- the approach to key facilities.

Pathways, hallways, doorways, stairways, access to floors above, and spaces to approach and reach essential facilities and controls in the home, are taken into consideration.

Principle 3: Adaptability

Adaptability means that a building or product can be simply adapted to meet a person's changing needs over time, or to suit the needs of different users. Many adaptations or adjustments within a Lifetime Home should be less disruptive, and more cost-effective, because the original design accommodates their future provision from the outset.

In a Lifetime Home, integral design and specification features, which may not be noticeable, will facilitate adaptations at a later stage for a household that has a family member with a temporary or permanent disability, or a progressive condition. A member of the household, or a visitor, will be able to live, sleep and use bathroom facilities solely on the entrance level for a short period, or the household can benefit from potential step-free access to upper-floor facilities.

Principle 4: Sustainability

Sustainability, in this context, refers to strong and stable communities underpinned by essential accessible elements aimed at meeting current and future needs, including homes, facilities, goods and services (the design of which contribute to the long-term viability of the neighbourhood and community).

The accessibility, flexibility and adaptability of a Lifetime Home all help to ensure long-term demand for, and desirability of, the dwelling. While sustainability is dependent on a range of factors, dwellings that offer this degree of accessibility and flexibility

are likely to remain popular over time, for both existing and new households, and can therefore contribute to the creation of thriving and popular neighbourhoods and communities.

Principle 5: Good value

Lifetime Homes are not intended to be complicated or expensive for house-builders or for the people who live in them. The design criteria have been carefully considered so that they can be incorporated into a dwelling's design and construction at building stage, with only a marginal cost effect.

Once occupied, the adaptability of the dwelling should actually save a household money if needs change and the dwelling is quickly and simply adapted to suit the new set of circumstances. Without the Lifetime Homes features, the household may be faced with expensive, complicated and disruptive major adaptation works to a dwelling less suited to change; or possibly (in the case of an existing household) face a forced move to a more suitable home.

Inclusive design and enabling simple adaptation from the outset has potential for considerable cost savings in the future, and long-term cost-effectiveness.

LIFETIME HOMES AND WHEELCHAIR STANDARD HOUSING

While a Lifetime Home may offer enough flexibility for some households with wheelchair-user occupants and regular visitors, the Lifetime Homes Standard[1] is quite different from the wheelchair housing design standard[2]. It also differs from other design standards relating to wheelchair users, such as the Greater London Authority (GLA) 2007 best practice guidance[3].

Wheelchair standard housing is considerably more detailed and demanding than the Lifetime Homes Standard in its spatial requirements and specifications. These higher requirements enable the entire property, and all its facilities, to be fully accessible and/or adaptable to suit many different wheelchair users. A wheelchair-adaptable standard, such as that produced by the GLA, provides spatial and structural requirements so that all areas and facilities within the entire dwelling can be fitted out to become fully accessible.

A wheelchair user living in a Lifetime Home will encounter particular accessibility and spatial compromises, depending on his or her circumstances. Some people will need, or prefer, the increased spatial and specification standards provided by wheelchair standard housing. It is vitally important that local planners and housing providers ensure that good provision of full wheelchair standard accommodation is made to meet this need.

LIFETIME HOMES AND APPROVED DOCUMENT M OF THE BUILDING REGULATIONS (ENGLAND & WALES)

The Lifetime Homes Standard was considered within a review of Approved Document (AD) M of the Building Regulations for England and Wales[4] in the late 1990s. At that time, AD M (dealing with reasonable provision to ensure that buildings are accessible and usable by people regardless of disability, age or gender) was being extended to include provisions for new dwellings. The resultant 1999 edition of AD M (and later editions) includes requirements equivalent to some, but not all, of the Lifetime Homes design and specification criteria. Those included in AD M tend to relate to two requirements:
• basic accessibility and 'visitability' into the entrance level of the dwelling
• access to service controls.

They exclude the Lifetime Homes provisions for future adaptability. A comparison of the Lifetime Homes Standard and AD M can be found on the Lifetime Homes web site (www.lifetimehomes.org.uk).

LIFETIME HOMES IN POLICY AND REGULATION

Some planning and funding authorities require newly built homes (or a proportion of new homes) to exceed the AD M requirements[4] and achieve the Lifetime Homes Standard[1]. As a leading example, the GLA adopted the Lifetime Homes Standard in the Supplementary Planning Guidance of the London Plan, issued in 2004[5]. This stated that all residential units in new housing developments should be Lifetime Homes, including houses and flats of varying sizes, in both the public and private sectors.

In 2008, the UK Government adopted Lifetime Homes in its report, *Lifetime Homes, Lifetime Neighbourhoods: A national strategy for housing in an ageing society*[6]. This announced targets for the building of all new housing to the Lifetime Homes Standard in both the public and private sectors. Following the change of government in the UK in 2010, these targets have not been upheld and responsibility for policy decision and implementation has passed to local authorities and their strategic partners.

Although not a regulatory body, the work of BSI is also relevant, as this work will inform and guide future regulation. In 2007, BSI published a British Standard Draft for Development (DD 266:2007) on the design of accessible housing[7]. This document built on, and made some suggested changes to the pre-July 2010 Lifetime Homes design criteria. The Draft is under consideration for conversion to a British Standard. This Draft document, while embracing the principles underlying the 2010 Lifetime Homes Standard, differs in points of detail from the Standard and should not be confused with it.

THE LIFETIME HOMES DESIGN CRITERIA

The 16 Lifetime Homes criteria relate to areas or features of the home and are listed in Table 1. For reference to the design criteria and other information and resources relating to Lifetime Homes see also the Lifetime Homes website, www.lifetimehomes.org.uk.

Table 1: Lifetime Homes design criteria

Criterion	Area or feature
1a	'On plot' (non-communal) parking
1b	Communal or shared parking
2	Approach to dwelling from parking
3	Approach to all entrances
4	Entrances
5	Communal stairs and lifts
6	Internal doorways and hallways
7	Circulation space
8	Entrance-level living space
9	Potential for entrance-level bed-space
10	Entrance-level WC and shower drainage
11	WC and bathroom walls
12	Stairs and potential through-floor lift in dwellings
13	Potential for fitting of hoists and bedroom/bathroom relationship
14	Bathrooms
15	Glazing and window handle heights
16	Location of service controls

2 FORMAT AND USE OF THE GUIDE

FORMAT

This Guide differs from other Lifetime Homes guidance in that its format does not follow a numerical pattern linked to the 16 Lifetime Homes design criteria (see Table 1).

The Guide has seven main technical design sections:

1 Approaching the home
2 Entrances
3 Internal circulation within communal areas
4 Entrance-level facilities within the home
5 Circulation and accessibility within the home
6 Circulation between storeys within the home
7 Service and ventilation controls.

Each section has an introduction describing the relevant Lifetime Homes objectives. The section is then divided into a number of headings and the relevant Lifetime Homes design criteria and/or specification requirements are incorporated into the guidance following each heading.

ADDITIONAL GOOD PRACTICE RECOMMENDATIONS

Also included are a number of 'additional good practice recommendations'. These are further design and/or specification good practice recommendations: they are not necessary for Lifetime Home compliance, but are consistent with the relevant inclusive design principles.

USE OF THE VERB 'SHOULD'

The use of the verb 'should' throughout this Guide indicates a Lifetime Homes design or specification requirement. For example, 'the surface should be non-slip' indicates that, for compliance, a non-slip surface is required. The verb 'must' is not used within the Guide to avoid confusion with other statutory obligations.

STATUTORY AND LEGISLATIVE REQUIREMENTS

It is a duty of designers referring to this Guide to ensure that all statutory and legislative requirements are met in full on any development. Statutory and legislative requirements current at the time of development must take precedence over the requirements given in this Guide.

TECHNICAL GUIDANCE

3   APPROACHING THE HOME

3.1   INTRODUCTION

Within the external environment the principal aim of the Lifetime Homes Standard[1] is to make the approach to dwellings, or blocks of dwellings, as convenient as practicable for as many people as possible. The Standard will particularly benefit parents with young children, people carrying shopping and those less agile or with reduced mobility.

Particular consideration is given to the ease with which people can get into and out of parked vehicles, and making movement between the parked vehicle and the dwelling as convenient as possible.

Where approaches to entrances are not from parked vehicles, the Standard also seeks to make these approach routes as convenient as is practicable, for as many people as possible.

At the main entrance, a Lifetime Home will provide people unlocking, or waiting at, the door some degree of shelter from the weather. The Standard also considers the ease of movement at, and through, the entrance doorway.

3.2   PARKING

3.2.1   Provision of parking

The Lifetime Homes Standard does not specifically require the provision of parking within developments. However, where parking is not provided, consultation with the local planning department is recommended to discuss local parking arrangements for any residents, or visitors, who may need to use a car. Where parking is provided, factors influencing its potential location, relating to a convenient and accessible approach route between it and the dwelling(s), as detailed in section 3.3, should be considered at the earliest planning stages.

3.2.2   Parking width for communal parking

Where parking is provided within communal or shared bays, at least one parking space with a minimum width of 3300 mm and a minimum depth of 4800 mm should be provided adjacent to (or close to) each block's entrance or (in the case of basement parking) lift core.

This space should be in addition to any designated spaces relating to any 'wheelchair housing' or 'wheelchair accessible housing' within the development.

The parking space with this width should be located as close to relevant entrances as practicable (see section 3.3.2).

3.2.3   Parking width for 'private' (on-plot) parking

If a dwelling has car parking provided within its individual plot or title boundary, at least one parking space length should be a minimum width of 3300 mm, or capable of achieving this width by simple adaptation of adjacent soft landscaping.

If a 2400 mm wide parking space has the 900 mm access path required by AD M

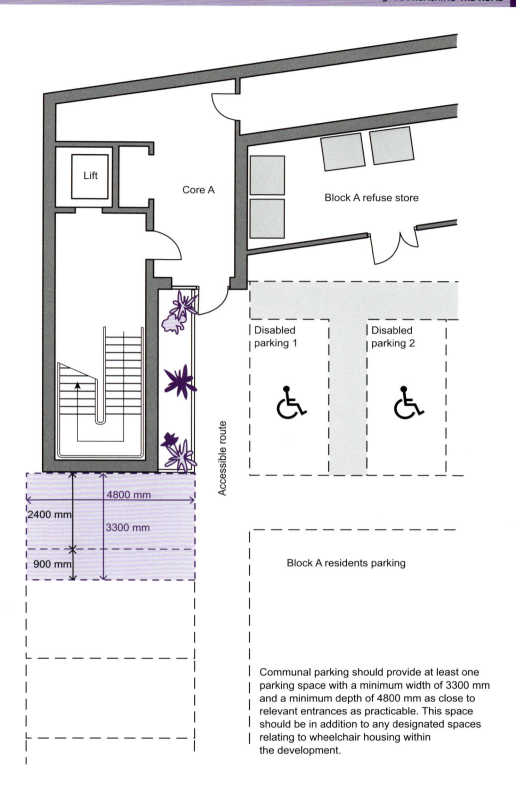

Lift

Core A

Block A refuse store

Disabled parking 1

Disabled parking 2

Accessible route

4800 mm

2400 mm

3300 mm

900 mm

Block A residents parking

Communal parking should provide at least one parking space with a minimum width of 3300 mm and a minimum depth of 4800 mm as close to relevant entrances as practicable. This space should be in addition to any designated spaces relating to wheelchair housing within the development.

Figure 1: **Parking width requirement for communal parking**

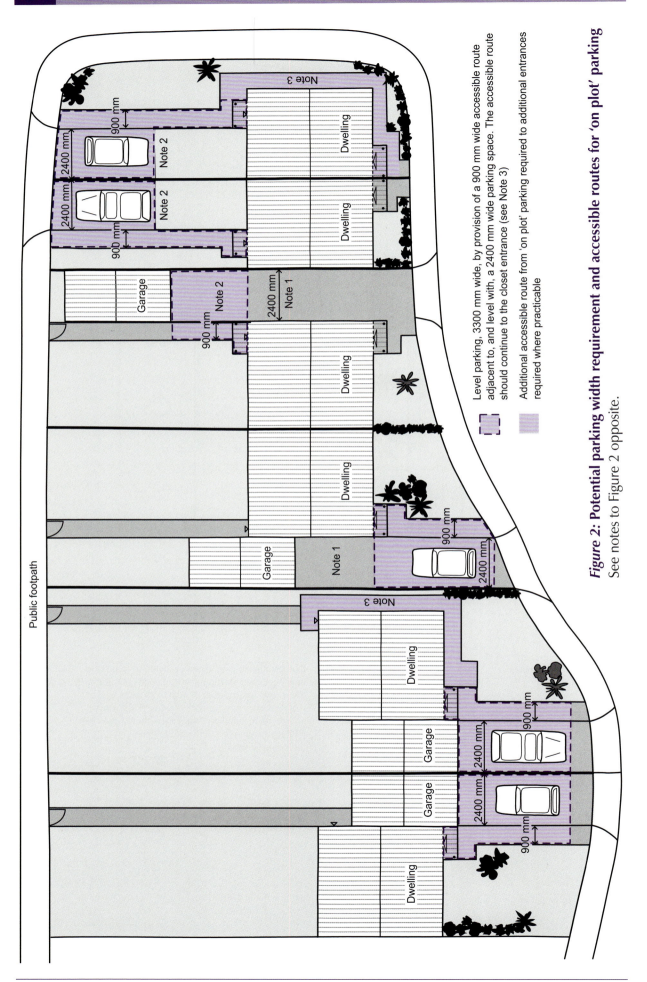

Level parking, 3300 mm wide, by provision of a 900 mm wide accessible route adjacent to, and level with, a 2400 mm wide parking space. The accessible route should continue to the closet entrance (see Note 3)

Additional accessible route from 'on plot' parking required to additional entrances required where practicable

***Figure 2:* Potential parking width requirement and accessible routes for 'on plot' parking**
See notes to Figure 2 opposite.

Notes to Figure 2:

1   If the only car hardstanding available is between fixed structures, then this would need to have a minimum width of 3300 mm.

2   Parking spaces providing the 3300 mm width/potential width should only be to the rear of dwellings where topography or regulation prevents them from being closer to the principal entrance.

3   Accessible routes (see section 3.3) from parking should preferably be to both the principal and any secondary entrance. Where this is not practicable, or when prevented by topography or regulation, the accessible route should be to/from the entrance closest to the parking space, and preferably to/from the principal entrance.

   A sole accessible route between parking and a secondary entrance should only arise where topography or regulation prevent an accessible route from the parking space to the principal entrance.

adjacent to it, and level with it, then the 3300 mm width requirement will automatically be met (Figure 2).

The location of this parking space in relation to the dwelling's entrances should take account of the requirements for an accessible route (and preferences expressed) discussed in section 3.3.

3.2.4 Gradients and surfaces of parking spaces

The 3300 mm wide spaces discussed above (whether present from the outset, or to be provided by widening) should have a firm and 'level' surface with no gradient exceeding 1:60, and no cross-fall for drainage exceeding 1:40.

These gradient and surface requirements are also recommended for as many other parking spaces as practicable.

3.2.5 Garages

The 3300 mm parking width requirements discussed above do not apply to garages. However, they would apply to any hard-standing for a parked car in front of a garage if the hard-standing was within a dwelling's title plot or boundary.

3.2.6 Car ports or other covered parking

The 3300 mm width/widening capability, and gradients and surfaces discussed above would apply to a car port or covered 'private' parking if it was the only parking space available for the dwelling.

Additional good practice recommendations exceeding the minimum requirements for parking are given in Box 1.

3.3 ACCESSIBLE ROUTE(S) BETWEEN PARKING AND DWELLINGS OR BLOCKS OF DWELLINGS

3.3.1 Provision of accessible route(s)

An accessible route, satisfying the requirements in sections 3.3.2–3.3.7, should be provided from the car parking spaces discussed above to relevant entrance(s) of dwellings or blocks of dwellings.

This accessible route is required whether the car parking is 'on plot' or communal.

Accessible routes should be the principal access route to the dwelling (or block of dwellings) from the associated parking.

On large developments where vehicular 'drop-off points' are also provided, the accessible route requirements would also apply between the 'drop-off point' and main entrance(s) to the relevant block(s).

Box 1: Parking — additional good practice recommendations

Increase width of parking spaces

- Where possible, increase the minimum width for parking spaces associated with this Standard from 3300 mm to 3600 mm.
- Provide all carports with a minimum clear width of 3300 mm (3600 mm preferred) regardless of whether or not they provide the only parking space for the dwelling.
- Provide garages (particularly those supplying the only parking for a dwelling) with a minimum internal width of 3300 mm (3600 mm preferred).

Increase length of parking spaces

- Increase the length of the wider spaces as much as practicable.
- Distinguish between parking spaces and an adjacent path width by either contrasting surface finishes for each (maintaining a level transition between the two), or provision of a contrasting strip of surface material between the two (laid flush with each adjoining surface).

Prevent obstacles

- In communal parking areas, when car parking bays give 'end-on' parking to a footpath, provide bollards (or equivalent measures) to prevent cars from overhanging access paths. Ensure that the bollards are sited beyond the edge of the relevant access route to prevent them from being an obstacle hazard to pedestrians. They should be a minimum of 900 mm high, not linked by chains or ropes, and should not have any protrusions.

3.3.2 Relevant entrances for accessible route(s)

Where the accessible route is to a block of dwellings it should lead to/from the block's main communal entrance. In the case of basement parking with lift access, the accessible route should lead to/from the entrance door to the lift core.

An accessible route from parking to an individual dwelling should preferably be provided to both the principal, and any secondary, entrance to the dwelling. Where topography or regulation prevents an accessible route to both the principal entrance door and the secondary entrance door, the route should be provided to/from the entrance door closest to the parking (ie the route most likely to be used by the household). Preferably, this should be between the parking and the dwelling's principal entrance.

A sole accessible route between the parking and a secondary entrance should occur only where it can be demonstrated that topography or regulation prevents the sole route being to the principal entrance (see Figure 2).

3.3.3 Distance of accessible route(s)

The parking spaces discussed in sections 3.2.2 and 3.2.3 should be located as close to relevant entrances as practicable, so that the accessible routes are as short as possible. For individual dwellings, the aim should therefore be to locate these parking spaces as close to the dwelling's main entrance as practicable.

Communal parking spaces discussed in section 3.2.2 should be as close as practicable to the relevant main entrance, or (in the case of basement parking) lift core(s), of the block(s) served. Other communal spaces should also be as close to the relevant entrances as practicable. On large developments with communal parking, all parking should preferably be within 50 m of associated entrances. If a distance in excess of 50 m cannot be avoided, level resting areas should be provided along the route.

3.3.4 Gradients along the accessible route(s)

The accessible route(s) should preferably be level (ie no gradient exceeding 1:60, and/or no crossfall exceeding 1:40). Where topography or regulation (eg in relation to flooding) prevents a level accessible route, it may be gently sloping in accordance with the maximum gradient/distance ratios set out in Figure 3 (ie a gradient of 1:12 for a distance of up to 2 m, and a gradient of 1:20 for a distance of 10 m, with gradients for intermediate distances interpolated between these values (eg 1:13 for a distance of 3 m, 1:14 for a distance of 4 metres). However, no single slope should have a going greater than 10 m in length (where necessary, intermediate level landings, at least 1.2 m in length should be provided to any distance in excess of 10 m).

All slopes (regardless of length) should have top and bottom landings, a minimum 1.2 m in length, clear of the swing of any door or gate.

Steps are not permitted at any point on the accessible route(s).

On sloping sites, the vehicular approach to parking space(s) should seek to manage the overall gradient so that the above requirements can be achieved between the parking space(s) and the relevant entrance(s).

3.3.5 Width of accessible route(s)

The accessible routes described above within individual dwelling plots and approaching individual dwellings should maintain a clear minimum width of 900 mm. Communal accessible routes should maintain a clear minimum width of 1200 mm.

3.3.6 Level landings at external entrance(s)

An external 'level' landing (ie with minimum crossfall to enable efficient drainage) should be provided adjacent to the entrance door at the end of the accessible route.

Similar 'level' landings should also be provided at any other main dwelling, or main communal, entrance (see section 4.7).

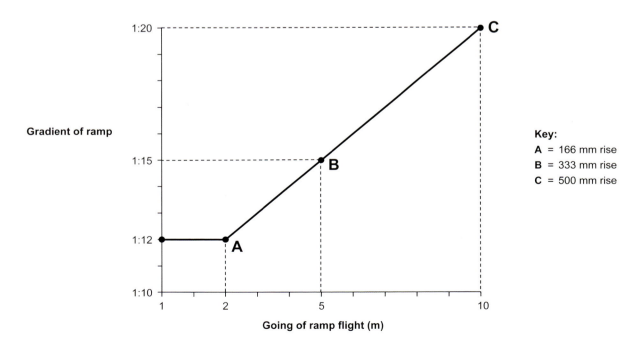

Figure 3: **Relationship between the gradient and going of a slope**

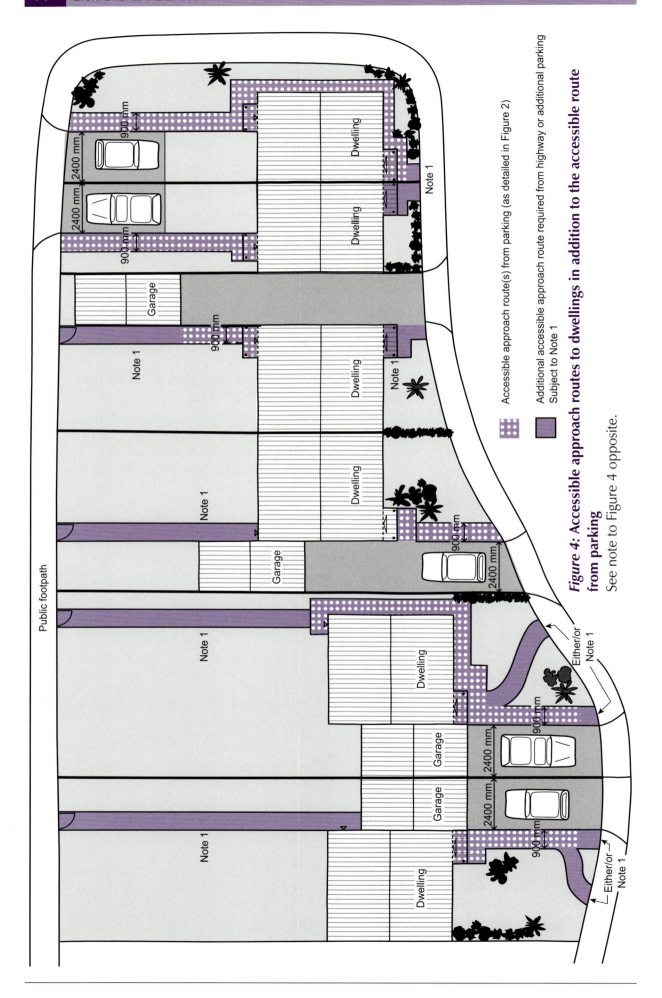

Figure 4: Accessible approach routes to dwellings in addition to the accessible route from parking

See note to Figure 4 opposite.

Accessible approach route(s) from parking (as detailed in Figure 2)

Additional accessible approach route required from highway or additional parking
Subject to Note 1

Note to Figure 4:

1   On steeply sloping sites the Standard recognises that a level, gently sloping approach on all paths to all entrances, may not be practicable or achievable. Where this is the case, the details should be discussed with the local planning authority to agree a workable solution. However, the Standard will always require the provision of an accessible approach route between parking and a relevant entrance (as Figures 1 and 2), regardless of topography.

At an individual dwelling's door, the minimum dimension for the 'level' landing should be 1200 mm × 1200 mm, clear of any door swing. At a communal entrance, the minimum dimensions should be 1500 mm × 1500 mm, clear of any door swing.

3.3.7 Surface of accessible route(s)

The surfaces of all accessible routes (and other approach routes – see section 3.4.3) should be firm, reasonably smooth and non-slip.

3.3.8 Additional stepped approach

If a principal accessible route to a communal entrance is not level and requires gentle slopes/ramps (in accordance with Figure 3), an additional, stepped approach route, in accordance with AD M (dwellings) requirements, should also be provided between the parking and relevant entrance(s).

Additional good practice recommendations exceeding the minimum requirements for accessible routes are given in Box 2.

3.4 OTHER APPROACH ROUTES TO DWELLINGS (FROM OTHER PARKING OR FROM THE HIGHWAY)

3.4.1 Gradients on other approach routes

All other approach routes to a dwelling entrance or a communal entrance from any parking or highway (in addition to the required accessible route(s) discussed in section 3.3) should (subject to the caveat below) be level or gently sloping, with gradient/distance ratios not exceeding the requirements detailed in section 3.3.4 and Figure 3.

However, the Standard recognises that, on steeply sloping sites, a level or gentle slope on all approaches to all entrances may not be practicable or achievable. Where this is the case, the details should be discussed with the local planning authority to agree a workable solution. However, the requirements for appropriate accessible routes between parking and relevant entrances, as detailed in sections 3.3.1–3.3.7, must not be compromised, and will always remain applicable if Lifetime Homes compliance is to be achieved.

3.4.2 Width on all approach paths

Minimum width requirements on all approach paths are consistent with the width requirements for the accessible approaches from parking detailed above, and should conform to section 3.3.5.

3.4.3 Surface treatments on all approach paths

Surface treatments on all approach paths are consistent with the surface requirements for accessible approaches from parking detailed above, and should conform to section 3.3.7.

Additional good practice recommendations exceeding the minimum requirements for other approach routes to dwellings are as given for accessible routes in Box 2.

Box 2: Accessible routes — additional good practice recommendations

Consider the need for setting-down points

- Provide large blocks of flats with a setting-down point close to the main entrance, with both a dropped kerb to allow easy access for wheelchair users from the road to the entrance and a kerbed section of footpath to enable ramped access to/from a taxi, with an accessible route to relevant entrances.

Minimise crossfalls

- Minimise crossfalls along all routes.

Increase width

- Where a 900 mm wide pathway route changes direction, increase the width in one direction to 1200 mm, or (preferably) maintain a clear width of 1200 mm for the entire length of all accessible routes.
- Provide a minimum width of 1800 mm on accessible routes that approach large blocks of flats.

Prevent obstacles

- Position bins, posts, bollards, outward opening windows and other potential obstacles beyond the route boundary and not overhanging the route for a height of 2250 mm.
- If providing gates on approach paths to individual dwellings ensure that they do not open outwards across any other path.

Provide clear opening widths for gates

- Provide clear opening widths at pedestrian gateways which are the same as the clear opening width of the entrance door associated with that approach route.

Minimise trip hazards

- Ensure that there is no difference in level at joints between paving units or that any unavoidable difference is no more than 5 mm. Joints should be finished flush, or, if recessed, no deeper than 5 mm. Joints between units should be no wider than 10 mm if filled or 5 mm if unfilled.
- Ensure that specifications for ground works and surface materials eliminate or minimise risk of differential settlement.
- Locate drainage gratings, inspection chamber covers, etc. off access routes wherever possible as wear and tear or differential settlement or both may cause them to become a trip hazard. They may also cause a barrier during maintenance and inspection.
- Where drainage gratings must be sited on access routes, ensure that their grill lines are perpendicular to the direction of travel to minimise the risk of wheelchair wheels or canes being disturbed or trapped.

Provide contrast

- Whenever possible, provide surface finishes on accessible routes that have tonal contrast with their adjacent surfaces. For routes that do not contrast with adjacent surfaces, provide a contrasting kerb edge that finishes flush with the level of the path surface.

Provide raised kerbs and handrails where ground slopes

- Provide a raised kerb and handrail or guarding for routes where adjacent ground falls away.

Provide consistent lighting levels

- Provide a consistent level of lighting of at least 50 lux on the route surface on communal access routes to blocks of dwellings.

3.5 OTHER EXTERNAL PATHS WITHIN DWELLING OR BLOCK CURTILAGES

The only external paths not covered by sections 3.3 and 3.4, are paths within dwelling gardens, or paths within semi-private communal areas of blocks of dwellings, that do not lead to either a parking space, or the highway (ie solely serving the private garden or communal semi-private area). There are no specific Lifetime Homes requirements for these paths, although including features listed in sections 3.3.4–3.3.7 (regarding gradients, widths, landings and surface treatments), as far as practicable, is recommended, particularly for any garden path leading to or from garden facilities (such as clothes lines, sheds, seating areas and outside taps).

4   ENTRANCES

4.1   INTRODUCTION

The Lifetime Homes Standard[1] considers the ease of movement at, and through, all entrance doorways, seeking to make this as convenient as possible for the widest range of people, including those pushing child buggies, those less agile, and those using sticks or other mobility aids.

At main entrances, a Lifetime Home will also provide some degree of shelter from the weather to people unlocking, or waiting at, the door.

4.2   EXTERNAL LIGHTING

4.2.1   Application

All entrances, whether to a private dwelling or a block of dwellings, should be lit externally with fully diffused luminaires. This requirement applies to any external door where a person may move across the threshold to or from an external space. This includes all balcony and roof terrace entrances (unless the balcony is a 'Juliet' type, where no access onto the balcony is intended).

4.3   ACCESSIBLE THRESHOLDS

4.3.1   Application

The requirement for an accessible threshold applies to all entrance and external doors to a dwelling and all communal entrance and external doors where a person may move across the threshold. Subject to the two exemptions stated below, all balcony doors and doors to all roof terraces (whether private or communal) should also meet this requirement.

Exemptions

Only 'Juliet' balconies (where no access onto the balcony is intended), and balconies or roof terraces over habitable rooms, which require a 'step-up' due to an increase in slab thickness, to provide additional thermal insulation to the accommodation below the balcony or roof terrace, are exempt.

4.3.2   Maximum up-stand

An accessible threshold should have an overall maximum up-stand of 15 mm. This 15 mm up-stand relates to the total height of the threshold unit (usually a one-piece proprietary product). In practice, it should consist of a number of smaller up-stands and sloping infill connections. Slopes on sills supporting the threshold, towards the external landing, should not exceed 15°.

4.3.3   Internal transition units

Where necessary, a small internal transition unit can be provided between the threshold and the internal floor finish. The maximum slope for any transition unit is also 15°.

Further guidance on, and examples of, accessible thresholds are given in the publication, *Accessible thresholds in new housing: guidance for house builders and designers*[8].

4.4   EFFECTIVE CLEAR OPENING WIDTHS OF ENTRANCE DOORS

The effective clear opening width of an entrance door is defined as:

the width of the opening measured in the same plane as the wall in which the door opening is situated, between a line perpendicular to the wall from the outside edge of the door stop on the latch side and the nearest obstruction on the hinge side when the door is open. The nearest obstruction may be projecting door furniture, a weatherboard, the door, or the door stop.

Figure 5 provides an illustration of effective clear opening width.

4.4.1 Dwelling entrance doors

All entrances to a dwelling (including all balcony and roof terrace doors) should have a minimum effective clear opening width of 800 mm, as detailed in Table 2.

4.4.2 Communal entrance doors

All communal entrances to/from any external area should have a minimum effective clear opening width of 800 mm or 825 mm depending on the direction and width of the approach to the door, as detailed in Table 2.

4.5   NIBS

4.5.1 Application

All entrance doors (without exception) should have a nib (or clear space) in the same plane as, and adjacent to, the door opening, at least 300 mm in length, provided to the leading edge, on the pull side, of the door (Figure 6).

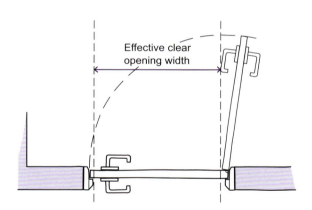 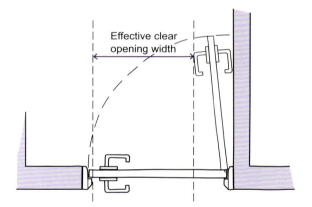

Figure 5: Effective clear opening width of entrance doors

Table 2: **Width of entrance doors**	
Direction and width of approach to door (external and internal)	**Minimum effective clear width (mm)**
Dwelling	800
Communal Straight on (without a turn or oblique approach)	800
At right angles to an access route: at least 1500 mm wide	800
at least 1200 mm wide	825

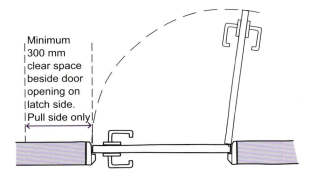

Minimum 300 mm clear space beside door opening on latch side. Pull side only

Figure 6: Nib requirement at all entrance doors

4.6 IRONMONGERY AND ACCESS CONTROLS

4.6.1 Height
Access controls (eg keyholes, locks, handles, door-entry system controls) should be no higher than 1200 mm from the ground level.

4.6.2 Location
Where there is a side or return wall or similar obstruction adjacent to the entrance, access controls (eg keyholes, locks, handles, door-entry system controls) should be sited at least 300 mm away from the corner between the entrance wall and the side/return wall.

An additional good practice recommendation exceeding the minimum requirements for ironmongery and access controls is given in Box 3.

Box 3: Ironmongery and access controls — additional good practice recommendation

- Provide a tonal contrast between all access controls, associated numbering/lettering and their backgrounds.

4.7 EXTERNAL LEVEL LANDINGS AT MAIN ENTRANCES
For the purposes of this guide, main entrances are defined as:

'the front door to an individual dwelling, the main communal entrance door to a block of dwellings, and any other entrance door associated with the accessible route detailed in sections 3.3.1 and 3.3.2.

4.7.1 Application
An external 'level' landing (ie with a minimum crossfall to enable efficient drainage) should be provided adjacent to all main entrance doors.

4.7.2 Dimensions
At a main entrance to an individual dwelling, the minimum dimension for the 'level' landing should be 1200 mm × 1200 mm, clear of any door swing.

At a main communal entrance the minimum dimensions should be 1500 mm × 1500 mm, clear of any door swing.

Additional good practice recommendations exceeding the minimum requirements for external level landings are given in Box 4.

Box 4: External level landings — additional good practice recommendations

- Provide level external landings of minimum dimensions 1200 mm × 1200 mm, clear of any door swing, at other secondary external entrances to a dwelling.
- Provide level external landings of minimum dimensions 1500 mm × 1500 mm, clear of any door swing, at all other external communal entrances to a block of dwellings (in addition to the main entrance).

4.8 WEATHER PROTECTION AT MAIN ENTRANCES

A definition of 'main entrances' is given in section 4.7.

4.8.1 Covers and canopies

Main entrances should have overhead cover to provide weather protection for those unlocking, or waiting at, the door.

4.8.2 Size and form of cover

The size and form of cover should be designed to take account of local conditions and the exposure at the relevant entrance, to ensure that effective weather protection is provided.

Cover at an individual dwelling's main entrance door should have a minimum depth of 600 mm (900 mm being typical), and cover at a main communal entrance door should have a minimum depth of 900 mm (1200 mm being typical). The width of the cover should exceed the width of the doorset plus any associated entry controls.

Additional cover and protection may be necessary at exposed sites.

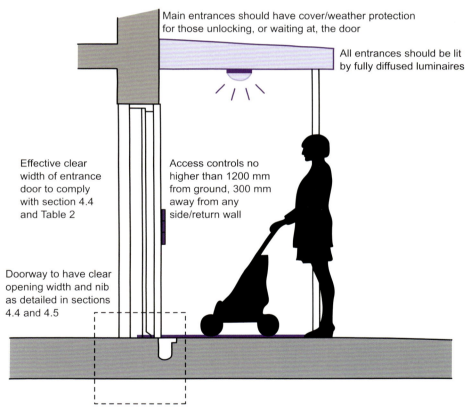

Main entrances should have cover/weather protection for those unlocking, or waiting at, the door

All entrances should be lit by fully diffused luminaires

Effective clear width of entrance door to comply with section 4.4 and Table 2

Access controls no higher than 1200 mm from ground, 300 mm away from any side/return wall

Doorway to have clear opening width and nib as detailed in sections 4.4 and 4.5

An accessible threshold is required at every entrance (including balcony and roof terraces unless the balcony or roof terrace is over habitable rooms and a 'step up' is required for thermal massing)

Figure 7: **Entrance requirements**

5 INTERNAL CIRCULATION WITHIN COMMUNAL AREAS

5.1 INTRODUCTION

The Lifetime Homes design criteria (Table 1) aim to provide sufficient clear widths within communal hallways and doorways to enable convenient movement through the doorways for the widest range of people (including those with pushchairs, those carrying children or objects, and those using wheelchairs or other mobility aids such as walking frames).

The width of the approach to a doorway should provide the necessary space to allow people using wheelchairs, or pushing buggies, to begin making any necessary turn into the doorway whilst approaching it, and the width of the door opening itself should enable any completion of the turn whilst passing through the opening. As a general principle, wider doorways will be required in narrower hallways.

In addition, sufficient space should also be provided to the side of the door to give people pushing child buggies, or using wheelchairs or walking frames, the necessary space to approach the door in a convenient manner whilst still being able to reach the ironmongery. This additional space to the side will also reduce the amount of subsequent movement and manœuvring necessary to be clear of the door swing.

The Lifetime Homes design criteria also seek to make vertical movement to storeys above or below the ground-floor/entrance level as convenient for as many people as possible. However, they recognise that maximum accessibility, enabled by the provision of passenger lifts, is not always financially viable in smaller blocks of flats.

5.2 INTERNAL COMMUNAL DOOR WIDTHS

The minimum clear opening width for all communal doorways with a 'head-on' (straight) approach route is 800 mm.

When the approach route to a communal doorway is not 'head-on', and a turn is required to pass through the doorway, the minimum clear opening width for the doorway should relate to the width of the approach route (typically a hallway, corridor or landing) as detailed in Table 3.

Table 3: Correlation between minimum communal hallway/corridor/landing widths and minimum clear opening widths of communal doors

Direction and width of approach	Minimum clear opening width (mm)
Straight-on (without a turn or oblique approach)	800
At right angles to a corridor or landing: at least 1500 mm wide	800
at least 1200 mm wide	825

5.3   NIBS

5.3.1 Provision

All communal internal doorways should have a minimum 300 mm nib (or clear space), in the same plane as the wall in which the door is situated, to the leading edge of the door, on the pull side (Figure 8).

Additional good practice recommendations exceeding the minimum requirements for nibs are given in Box 5.

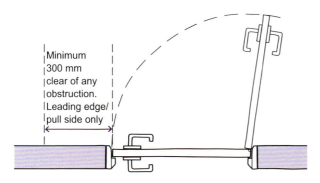

Minimum
300 mm
clear of any
obstruction.
Leading edge/
pull side only

Figure 8: **Nib requirement at all communal doors**

Box 5: **Nibs — additional good practice recommendations**

- In addition to the 300 mm nib required to the leading edge/pull side, provide a nib, to the push side, at least 200 mm in length, between the latch side of the door and any obstruction.
- Extend the length of the nibs beyond the minimum required to further assist by increasing the general approach and manoeuvring space around the door.

5.4   COMMUNAL HALLWAY, CORRIDOR AND LANDING WIDTHS

The minimum width for a communal hallway, corridor or landing will usually depend on the clear opening width of doorways situated with that hallway, corridor or landing (Figure 9).

If doorways in the side walls of a hallway, corridor or landing have a clear opening width of at least 825 mm, then the minimum width of that hallway, corridor or landing is 1200 mm. This may reduce to a minimum 1050 mm at 'pinch points' (eg beside a narrow structural column), as long as the reduced width is not opposite, or adjacent to, any doorway or change of direction within the hallway, corridor or landing.

However, if any doorways in the side walls of the hallway, corridor or landing have the minimum permitted clear opening width of 800 mm, then the hallway, corridor or landing should have a minimum width of 1500 mm. This may also reduce to a minimum 1050 mm at 'pinch points' (eg beside a narrow structural column), as long as the reduced width is not opposite, or adjacent to, any doorway or change of direction within the hallway, corridor or landing.

Additional good practice recommendations exceeding the minimum requirements for communal hallway, corridor and landing widths are given in Box 6.

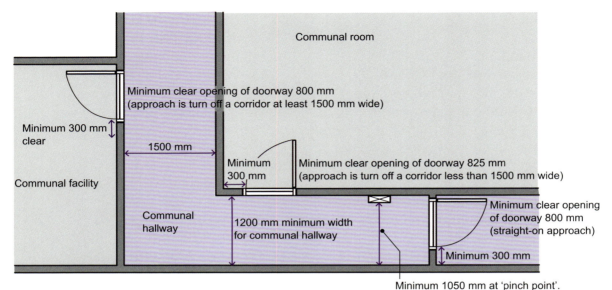

Figure 9: **Communal hallway and doorway widths**

Box 6: **Communal hallway, corridor and landing widths — additional good practice recommendations**

Increase width on high traffic routes

- In hallways, corridors or landings serving a high number of dwellings (eg on the entrance storey en-route to lifts) increase this width to 1500 mm or 1800 mm.

Enable easy manœuvring outside dwelling entrance doors

- Provide 1500 mm × 1500 mm turning points in communal hallways at, or adjacent to, all dwelling entrance doors.

Mats and mat wells

- There are no specific Lifetime Homes requirements relating to mats and mat

wells. However, where provided, they should enable the matting surface to be flush with the adjacent floor finish. Surface laid mats, if provided, should have chamfered edges that are not a vertical up-stand barrier or trip hazard. All matting should be of materials that are resistant to fraying and the pile should be suitably dense and shallow to minimise resistance to wheelchair movement. Matting should extend sufficiently into the building to clean the full circumference of wheelchair wheels and, where practicable, extend across the full width of the entrance area.

5.5 COMMUNAL STAIRS

5.5.1 Pitch

Communal staircases providing a principal access route to a dwelling should be 'easy going' with maximum uniform risers of 170 mm and minimum uniform goings of 250 mm.

5.5.2 Handrails

Handrails should be 900 mm above each nosing and they should extend 300 mm beyond the top and bottom step (Figure 10).

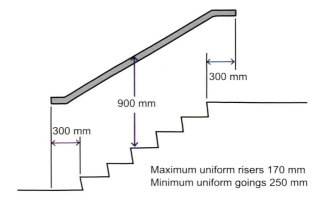

Maximum uniform risers 170 mm
Minimum uniform goings 250 mm

Figure 10: **Communal stairs**

5.5.3 Application

The above requirements apply to communal stairs within a block that provide a principal access route to a dwelling, regardless of whether or not a lift is provided.

Additional good practice recommendations exceeding the minimum requirements for communal stairs are given in Box 7.

Box 7: Communal stairs — additional good practice recommendations

- Provide contrasting surfaces to risers and treads and provide further differentiation to the nosing of each stair tread by using a contrasting strip.
- Provide handrails with a profile suitable for firm grip (eg circular mopstick), 45–50 mm in diameter, continuous on both sides of the stair. Handrail ends should be adequately end-stopped or turned in towards the wall to avoid the risk of clothing from snagging. Handrail supports and turns should not interfere with sliding hand movement/grip along the rail.
- Position prominent floor numbers, with tactile indicators, contrasting against their background, at consistent locations at the top and bottom of every flight.
- Design communal stairs to preclude the possibility of walking into the underside of any stairway.

5.6   COMMUNAL PASSENGER LIFTS

5.6.1 Provision of lifts and application

There is no Lifetime Homes requirement for the provision of a communal passenger lift within communal hallways. However, where lifts are provided, the requirements listed below would apply to every lift within each block.

5.6.2 Size of lift

Where passenger lifts are provided, their minimum internal floor dimensions should be 1100 mm × 1400 mm.

5.6.3 Position of lift controls

Where a lift is provided, the controls should be between 900 mm and 1200 mm from the floor and 400 mm from the lift's internal front wall.

5.6.4 Landings

A clear landing of 1500 mm × 1500 mm should be provided on each storey adjacent to all lifts.

Additional good practice recommendations exceeding the minimum requirements for communal passenger lifts are given in Box 8.

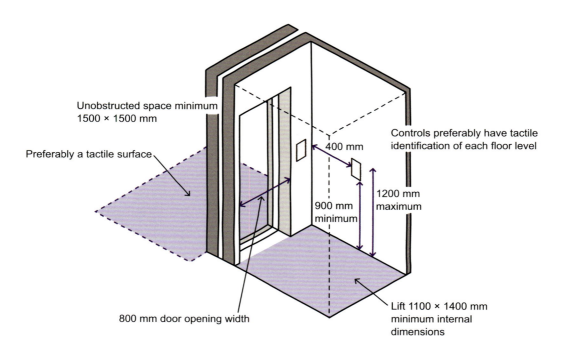

Figure 11: **Communal passenger lift**

Box 8: Communal passenger lifts — additional good practice recommendations

- Whenever practicable, provide lifts in accordance with the requirements given in sections 5.6.2–5.6.4 to all dwellings situated above the entrance level of a block.
- In small blocks, where this is not financially viable, provide structure, space and services to enable the provision of a lift in the future.
- In blocks with four or more storeys, provide access to two lifts.
- Where lifts are provided, install at lift landings a change in contrast and texture to the floor finish to differentiate them from adjacent floor finishes. Also provide a completely flush and seamless transition between the two different colours/textures.
- Provide a floor finish that contrasts with the lift car walls, preferably of a lighter tone, to avoid the appearance of an open lift shaft to people with sight loss.
- Provide large and contrasting tactile controls for each lift.
- Provide audible as well as visual signs/ signals for the movement, action and location of any lift.

6   ENTRANCE-LEVEL FACILITIES WITHIN THE HOME

6.1   INTRODUCTION

The Lifetime Homes Standard[1] requires a number of facilities to be provided on the entrance level of each dwelling.

Some of these entrance-level facilities, together with the accessibility principles on the approach to the dwelling (as detailed in preceding sections), combine to make visiting the dwelling as convenient as practicable for people who are less able to use stairs.

Provision of these entrance-level facilities will also enable a household to make simple and cost-effective adjustments to the dwelling to enable a member of the household temporarily to live and sleep on the entrance level if they are unable to use the stairs for a period of time (eg after a hip operation). Other requirements, as detailed later in chapter 8, can provide more permanent solutions for access to the upper storey(s) when a member of the household is unable to use the stairs.

6.1.1 Definition of 'entrance level'

For the purposes of Lifetime Homes design and specification, the entrance level of a dwelling is generally defined as follows.

- For a dwelling with a direct external entrance, the entrance level is:

the storey containing an entrance door approached by an 'accessible route' as detailed in section 3.3.

- For a dwelling that is approached via a communal hall, landing, stair or lift, the entrance level is:

the storey of the block containing the entrance door of that particular dwelling.

- For a dwelling that has no habitable or non-habitable rooms on the storey containing its entrance door(s) (eg most flats over garages, some flats over shops, some duplexes and some townhouses), the entrance level is:

the first-storey level containing a habitable or non-habitable room if this storey is reached by an 'easy going' stair with maximum risers 170 mm, minimum goings 250 mm, and a minimum width of 900 mm measured 450 mm above the pitch line.

6.2   LIVING SPACE

The 'living space' is defined as:

any permanent living room, living area, dining room, dining area (possibly within a kitchen diner), or other permanent reception area that provides seating and socialising space for the household and visitors.

6.2.1 Provision

A living room or living space should be provided on the entrance level of every dwelling.

6.2.2 Sight line through windows of principal living spaces

Where the entrance-level living space is also the principal living space (typically the living room) its principal window or glazed doors (where these are in lieu of a principal window) should have glazing that starts no higher than 800 mm above the finished floor level to allow seated people a view through the window (Figure 12). In addition, any full width transom or sill of the principal window within the field of vision (extending up to 1700 mm above floor level) should be at least 400 mm in height away from any other transom or non-transparent balcony balustrade. The above requirements should also apply to any other living space deemed to be the principal living space. All dimensional requirements within this paragraph have a 50 mm tolerance.

6.2.3 Temporary bed-space and potential through-floor lift

It should be noted that the entrance-level living space may also need to provide for the temporary bed-space (see section 6.3) and potential position for a through-floor lift (see section 8.3) unless these are provided elsewhere on the entrance level.

An additional good practice recommendation exceeding the minimum requirements for entrance-level living space is given in Box 9.

Box 9: Entrance-level living space — additional good practice recommendation

Glazing line of windows of all living spaces

- Where the entrance-level living space is not the principal living space, or where there are additional living rooms/spaces, it is preferable to provide windows to these rooms/spaces with glazing lines that start no higher than 800 mm above finished floor level, and other requirements detailed in section 6.2.2, to allow seated people to have a view through the window(s).

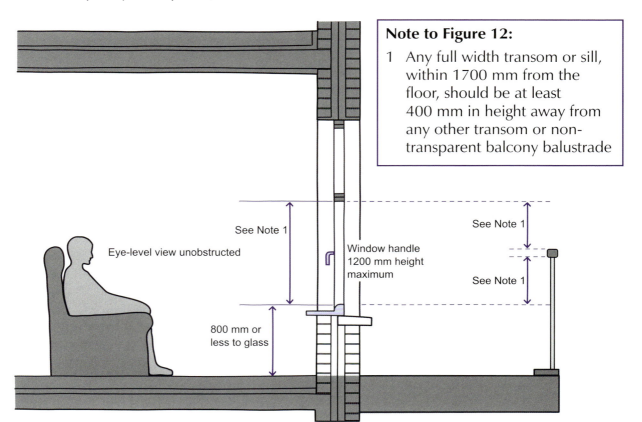

Note to Figure 12:

1 Any full width transom or sill, within 1700 mm from the floor, should be at least 400 mm in height away from any other transom or non-transparent balcony balustrade

See Note 1

Eye-level view unobstructed

Window handle 1200 mm height maximum

See Note 1

See Note 1

800 mm or less to glass

Figure 12: **Sight line through window of principal living space**

6.3 POTENTIAL FOR AN ENTRANCE-LEVEL BED-SPACE

6.3.1 Provision

Where there is no permanent bedroom on the entrance level, there should be space on the entrance level that could be used as a convenient temporary bed-space (Figure 13).

This space can be provided within a living room (eg in a corner of the room) following rearrangement of the room's furniture. However, the room, when containing the temporary bed-space, should also remain functional as a living room, despite its compromised layout.

Any other 'living space' as defined in section 6.2 can also provide for the

Example 1: Ground-floor plan without (Left) and with (Right) temporary bed-space

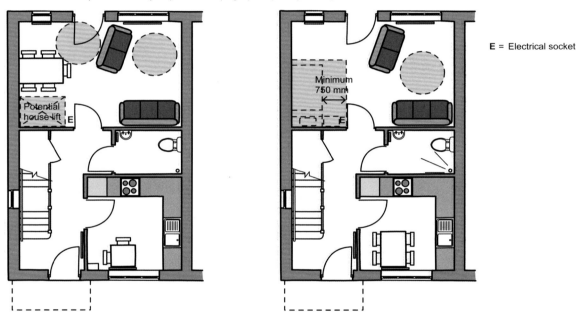

Example 2: Ground-floor plan without (Left) and with (Right) temporary bed-space

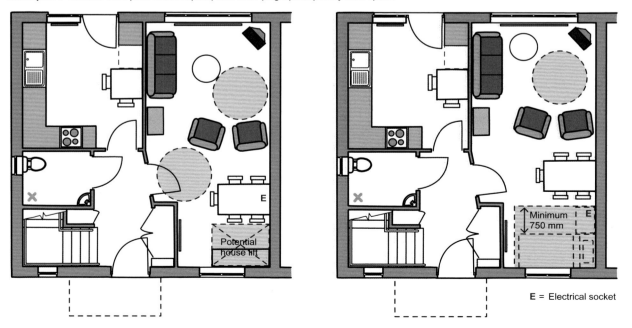

Figure 13: **Plans for potential entrance-level bed-space: Examples 1 and 2**

Example 3: Ground-floor plan without (Left) and with (Right) temporary bed-space

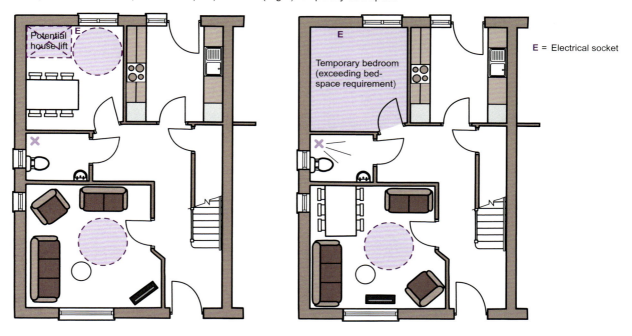

E = Electrical socket

Note: Examples 1–3 all demonstrate recommendations for heat source and window within potential bed-space area, and potential for efficient screening

Figure 13 (cont'd): **Plans for potential entrance-level bed-space: Example 3**

temporary bed-space. However, where the bed-space is provided within a dining room or dining area, the dining function must also be able to continue in the room or space, unless it can be relocated elsewhere.

Providing for this facility within a kitchen/diner would result in the least convenient arrangement and should be avoided whenever possible.

The space should be capable of being screened (with a portable screen) from the remainder of the room.

6.3.2 Size of space
The space required should be able to accommodate a single bed plus a 750 mm wide space to one side of the bed.

6.3.3 Electrical socket
The space should contain at least one electrical socket.

6.3.4 Overlap with potential 'through-floor' lift position
The potential bed-space can overlap with the identified route for the potential 'through-floor' lift (see section 8.3), as the temporary bed-space would not be required if a through-floor lift enabled access to sleeping accommodation on a floor above (Figure 13).

Additional good practice recommendations exceeding the minimum requirements for potential for an entrance-level bed-space are given in Box 10.

Box 10: Potential for an entrance-level bed-space — additional good practice recommendations

- Design a layout that provides a suitable recess or portion of the room for the bed-space. Enabling better separation between it and the remainder of the room is desirable.
- Provide a window for ventilation and a heat source with an independent heat control to create additional comfort within the area.

6.4 PROVISION OF AN ACCESSIBLE WC, BASIN AND SHOWER DRAINAGE

6.4.1 Provision in two-bedroom houses and maisonettes (as defined below)

In dwellings with two or more storeys, and no more than two habitable rooms in addition to the main living room and any kitchen diner (typically a one- or two-bedroom house), provision of a WC on the entrance level that complies with AD M[4] will spatially satisfy this requirement. However, a floor drain for a future accessible shower (which is not required by AD M), as detailed in section 6.4.5, should be provided within the WC compartment, or in an entrance-level bathroom, or in a suitable location elsewhere on the entrance level.

6.4.2 Provision in all other dwellings

In all other dwellings, an accessible WC, basin and drainage for a potential accessible shower, as detailed in sections 6.4.3–6.4.5, should be located on the entrance level.

These may be provided within an accessible bathroom on the entrance level (see section 7.6.2), or within a separate WC compartment/cloakroom on the entrance level.

6.4.3 Accessible WC

An accessible WC (Figure 14) should be provided with:

- a centre-line between 400 mm and 500 mm from an adjacent wall
- a flush control located between the centre-line of the WC and the side of the cistern furthest away from the adjacent wall
- an approach zone extending at least 350 mm from the WC's centre-line towards the adjacent wall, and at least 1000 mm from the WC's centre-line on the other side. This zone should extend forward from the front rim of the WC by at least 1100 mm. The zone should also extend back at least 500 mm from the front rim of the WC for a width of 1000 mm from the WC's centre-line
- an accessible basin, as detailed in section 6.4.4, which may be located either on the adjacent wall, or adjacent to the cistern, but should not project into this approach zone by more than 200 mm.

6.4.4 Accessible basin

An accessible basin (Figure 15) should be provided with a clear frontal approach zone extending back for a distance of 1100 mm from any obstruction under the basin such as a pedestal, trap, duct or housing. This zone will normally overlap with the WC's approach zone as detailed in section 6.4.3.

6.4.5 Floor drainage for an accessible shower

Unless provided elsewhere on the entrance level, the entrance-level accessible bathroom or the entrance-level accessible WC compartment should also contain floor drainage for an accessible floor-level shower.

A floor construction is required that either provides shallow falls to floor drainage from the outset, or, where the drainage is initially capped for use later following installation of a shower, that allows simple and easy installation of a laid-to-fall floor

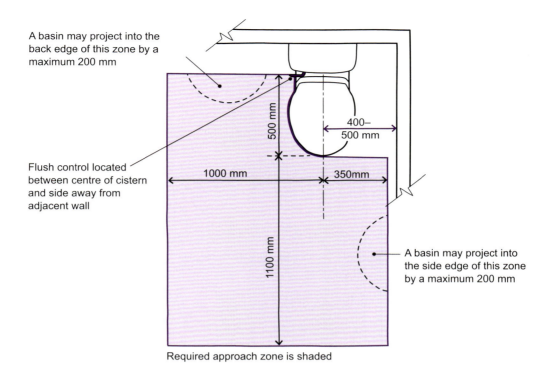

A basin may project into the back edge of this zone by a maximum 200 mm

500 mm

400–500 mm

Flush control located between centre of cistern and side away from adjacent wall

1000 mm   350mm

1100 mm

A basin may project into the side edge of this zone by a maximum 200 mm

Required approach zone is shaded

Figure 14: **Accessible WC**

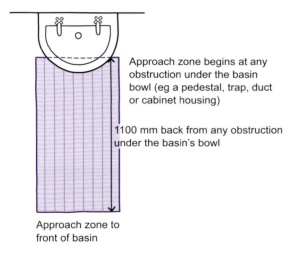

Approach zone begins at any obstruction under the basin bowl (eg a pedestal, trap, duct or cabinet housing)

1100 mm back from any obstruction under the basin's bowl

Approach zone to front of basin

Figure 15: **Accessible basin**

surface to the capped drainage in the future.

Any falls in the floor to the drainage, whether provided from the outset, or provided later by adaptation, should be the minimum required for efficient drainage of the showering catchment area. Crossfalls in the floor should be minimised.

The drainage should be located as far away from the room's doorway as practicable.

The requirements of sections 6.4.3–6.4.5 are applicable within a WC compartment and are illustrated in Figure 16.

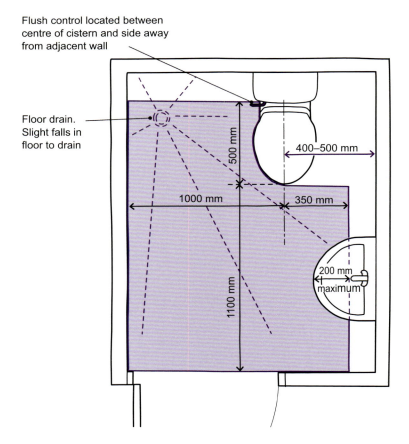

Figure 16: **Example of an accessible WC compartment**

6.4.6 Wall construction

To assist with independent use of facilities, all walls within the entrance-level bathroom or entrance-level WC compartment (and walls within all other WC or bathroom facilities throughout the dwelling) should be capable of immediate firm fixing and support for adaptations such as grab rails within a height band of 300 mm–1800 mm from the floor, as and when these become required (with no strengthening or adaptation to the wall structure being necessary).

An additional good practice recommendation exceeding the minimum requirements for entrance-level facilities within dwellings is given in Box 11.

> **Box 11: Entrance-level facilities in dwellings — additional good practice recommendation**
>
> • Provide a kitchen on the entrance level

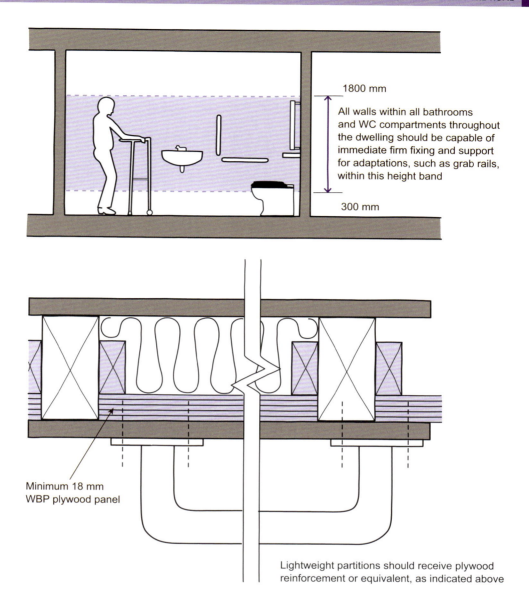

1800 mm

All walls within all bathrooms and WC compartments throughout the dwelling should be capable of immediate firm fixing and support for adaptations, such as grab rails, within this height band

300 mm

Minimum 18 mm WBP plywood panel

Lightweight partitions should receive plywood reinforcement or equivalent, as indicated above

Figure 17: **Wall construction for all bathrooms and WCs WBP = weather and boil proof**

7 CIRCULATION AND ACCESSIBILITY WITHIN THE HOME

7.1 INTRODUCTION

Movement in hallways, through doorways, and within rooms should be as convenient as possible to the widest range of people, including those using walking frames, wheelchairs or other mobility aids, and those moving furniture or other objects.

Key facilities within the dwelling should also be convenient to approach, and use, for the widest range of potential occupants.

The minimum circulation and approach spaces required within this chapter, whilst considering basic circulation space for a wheelchair user, do not match the equivalent space requirements within wheelchair housing, or wheelchair-adaptable housing. A wheelchair user living in a Lifetime Home will therefore need to accept compromises in respect of available manœuvring and circulation space. Some wheelchair users will choose, or need, the increased spatial arrangements, and/or the more specific component specification, provided (or enabled) by the wheelchair housing design standard[2] and the wheelchair-adaptable design standard[3].

However, consideration of the basic circulation space for wheelchair use, does create circulation and approach spaces that will assist a wide range of occupants, potential occupants, and visitors, including families with young children, or people using walking sticks or frames, and some wheelchair users.

The Lifetime Homes Standard[1] does not itself set overall minimum space standards for individual rooms or overall dwelling footprints, but relies on the various requirements at key circulation and approach spaces, as detailed in this chapter, to set a minimum standard for necessary space within these key areas. When planned and considered from the outset of design, these spatial requirements need not be onerous, and can often be incorporated with little effect on the overall dwelling area. However, it goes without saying that dwellings with decent space standards will enable easy incorporation of the requirements, and will offer more convenience to the occupiers.

Consideration is also given to enabling effective and efficient adaptation works to allow access to storeys above the entrance level for household members with less agility and mobility. Early layout planning should also therefore consider the possible future need for a convenient and simple route for a through-floor lift (as detailed in section 8.3).

The need for a convenient route between a main bedroom and bathroom as detailed in section 7.5.2 should also have early consideration.

Overall, the route from the entrance level via the potential through-floor lift to the storey above, subsequent access to a main bedroom and the route to the accessible bathroom from the relevant bedroom, should be as simple and convenient as possible.

7.2 INTERNAL HALLWAYS, LANDINGS AND DOORWAYS

A general principle to note in the early layout stages is that narrower hallways and landings will need wider door openings in their side walls. This is discussed in more detail in sections 7.2.1 and 7.2.2 and illustrated in Figure 18.

7.2.1 Minimum hallway and landing widths

The minimum width of any hallway or landing in a dwelling should be 900 mm, 1050 mm or 1200 mm, depending on the width and location of associated door openings. If a hallway or landing has no doorways in its side walls, its minimum width should be 900 mm. When a person in a hallway or on a landing needs to turn to pass through any associated doorway, the minimum width of that hallway (or landing) will depend on the minimum clear opening width available at those doorways — as detailed in Table 4. The minimum width of a hallway or landing may reduce to 750 mm at 'pinch points' (eg beside a radiator) as long as the reduced width is not opposite, or adjacent to, a doorway.

7.2.2 Minimum internal doorway widths

The following clear opening width requirements apply to all doorways within dwellings where movement through the doorway is intended. They do not apply to storage/cupboard doors unless a person would need to enter the cupboard to access the storage.

When the approach route to the door is 'head-on', the minimum clear opening width of that doorway is 750 mm.

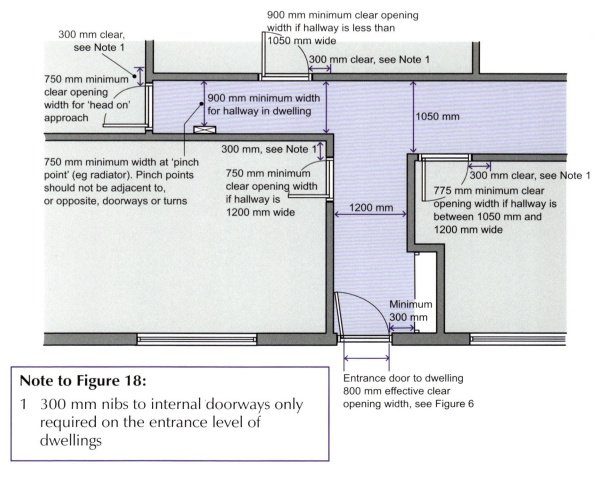

Note to Figure 18:

1   300 mm nibs to internal doorways only required on the entrance level of dwellings

Figure 18: **Hallways, landings and doorway widths in dwellings**

Table 4: Width of internal dwelling doors

Direction and width of approach (typically a hallway or landing)	Minimum clear opening width (mm)
Straight-on (without a turn or oblique approach)	750
At right angles to a hallway or landing	
at least 1200 mm wide	750
at least 1050 mm wide	775
less than 1050 mm wide (minimum width 900 mm)	900

When the approach to a doorway is not head-on, and a turn is required to pass through the doorway, the minimum clear opening width of that doorway will relate to the width of the approach (typically a hallway or landing), and should be in accordance with Table 4.

7.2.3 Nibs

This nib requirement is only applicable to doorways to rooms on the entrance level of each dwelling.

All doors to rooms on the entrance level of each dwelling should have a minimum 300 mm nib (or clear space) in the same plane as the wall in which the door is situated, to the leading edge of the door, on the pull side (Figure 19).

Nib requirements at entrance doors are detailed in section 4.5.

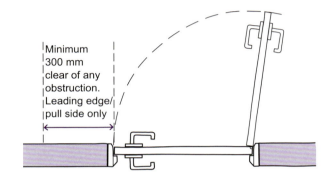

Minimum 300 mm clear of any obstruction. Leading edge/pull side only

Figure 19: **Nib requirement to room doorways on the entrance level of dwellings**

7.3   LIVING ROOMS/AREAS AND DINING ROOMS/AREAS

7.3.1 Turning circle or ellipse

Living rooms/areas and dining rooms/areas should be capable of having either a clear turning circle of 1500 mm diameter, or a turning ellipse of 1700 mm x 1400 mm (Figure 20). Occasional items of furniture that are easily moved (typically coffee tables and side tables) can be within, or overlap, these turning zones.

7.3.2 Circulation space between items of furniture

Where movement between furniture would be necessary for essential circulation (eg to approach doorways, other rooms, the window, or essential day-to-day controls, a clear width of 750 mm between furniture items along the necessary route(s) should be possible (Figure 20).

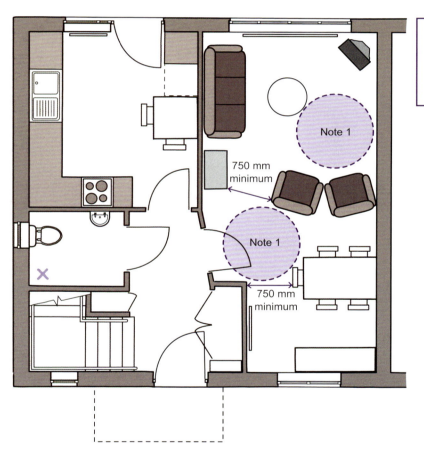

> **Note to Figure 20:**
>
> 1   1500 mm manœuvring circle or 1400 mm × 1700 mm ellipse

Figure 20: **Turning circle/ellipse and circulation space in living rooms and dining areas**

7.4 KITCHENS

7.4.1 Space in front of kitchen units and appliances

A clear width of 1200 mm should be provided between the fronts of kitchen units and appliances and any fixed obstruction opposite (such as other kitchen fittings or walls). This clear 1200 mm should be maintained for the entire run of the unit, worktop and/or appliance (Figure 21).

Additional good practice recommendations exceeding the minimum requirements for kitchens are given in Box 12.

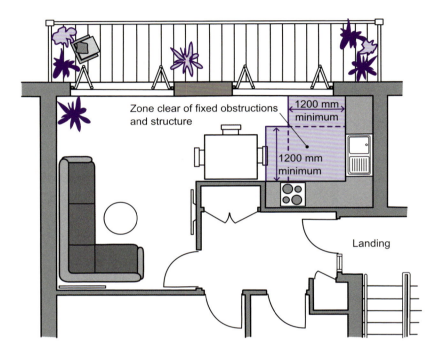

Figure 21: **1200 mm wide accessibility zone from kitchen units and appliances**

Box 12: Kitchens — additional good practice recommendations

- Plan kitchen layouts, whenever possible, so that they include, or can be re-fitted to provide, a continuous run of units unbroken by doorways, that includes:
  - a built-in oven at an accessible height beside a minimum 600 mm width of work surface
  - a hob beside a further minimum 600 mm width of work surface
  - a sink/drainer.

  This continuous run, uninterrupted by doorways (approximately 3600 mm in length measured along the front face) could be straight, L-shaped, or U-shaped. Space for other typical 'white goods' and fittings can be available elsewhere in the kitchen (only the oven and hob need to be contained within this particular length of run). Windows should not be located behind the hob.

- Provide a clear 1500 mm diameter circular, or 1400 mm × 1700 mm elliptical, manoeuvring space between floor level and a minimum height of 900 mm, for the benefit of wheelchair users.

7.5   BEDROOMS

A 'main bedroom' is defined here as:

any double or twin bedroom.

7.5.1 Circulation space within a main bedroom

A clear space, minimum 750 mm wide, should be possible to both sides and the foot of a standard-sized double bed within at least one main bedroom (Figure 22). Bedside cabinets, or other furniture, can be placed within these spaces adjacent to

the head of a bed, for a maximum depth of 600 mm.

7.5.2 Relationship of a main bedroom to the accessible bathroom

A main bedroom should be situated on the same storey as, and close to, an accessible bathroom. See section 7.6.2 for the requirements within an accessible bathroom. The route between this bedroom and bathroom should not pass through any living/habitable room or area. It is desirable, and typical, that

Main bedroom example 1

750 mm · 750 mm · 750 mm

Main bedroom example 2

750 mm · 750 mm · 750 mm

Note: Beside cabinets (or other furniture, maximum 600 mm depth) can be placed beside the head of beds within the 750 mm wide circulation spaces

Layouts above show examples of main bedroom layouts providing the minimum 750 mm clear around both sides and foot of bed

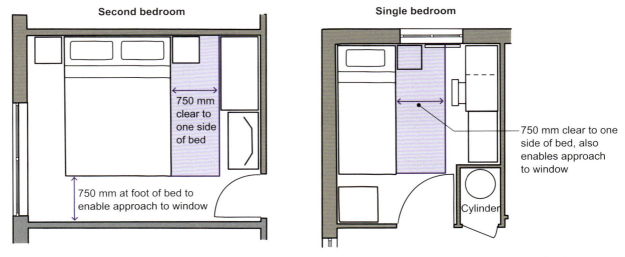

Second bedroom

750 mm clear to one side of bed

750 mm at foot of bed to enable approach to window

Single bedroom

750 mm clear to one side of bed, also enables approach to window

Cylinder

Layouts above show examples of other (non-main) bedrooms providing the minimum 750 mm clear to one side of bed. Note required clear 750 mm at foot of bed if it is necessary to pass it to approach the window

***Figure 22:* Bedroom circulation spaces**

this main bedroom should also contain the circulation spaces as detailed in section 7.5.1.

Potential for through-floor lift access

The storey containing this main bedroom and accessible bathroom should have potential access from the through-floor lift as detailed in section 8.3.

Structure over ceilings

The structure above the ceilings of this main bedroom and accessible bathroom should be capable of supporting, or capable of adaptation to support, the future installation of single point (static) hoists above the bed, bath and WC positions.

7.5.3 Circulation space within other bedrooms

Other bedrooms should be capable of having a clear space, minimum 750 mm wide, to one side of the bed (Figure 22). Bedside cabinets, or other furniture, can be placed within these spaces beside the head of a bed, for a maximum depth of 600 mm.

Approach to window past foot of bed

Where it is necessary to pass the foot of the bed in order to approach the window, a clear width of 750 mm should also be provided at the foot of the bed.

Additional good practice recommendations exceeding the minimum requirements for bedrooms are given in Box 13.

7.6   BATHROOMS AND WCs

7.6.1 Entrance-level accessible WC and shower drainage

An entrance-level WC and shower drainage, in accordance with sections 6.4.3–6.4.6, should be provided in each dwelling. This may be within an entrance-level WC compartment, or within an entrance-level bathroom.

The design and specification requirements for this are detailed in chapter 6.

7.6.2 Accessible bathroom

An accessible bathroom should be provided in every dwelling. In single-storey dwellings (flats and bungalows) this accessible bathroom will provide for the required entrance-level accessible WC and shower drainage, as detailed in section 7.6.1, and in sections 6.4.3–6.4.6.

The accessible bathroom should be close to, and on the same storey level as, a main (double or twin) bedroom. In dwellings with more than one storey (unless on the entrance level with a main bedroom) it should be on a storey level with potential access by a through-floor lift (see section 8.3).

The accessible bathroom should contain the following facilities listed 1–7, and associated clear approach zones.

Box 13: Bedrooms — additional good practice recommendations

- Position the main bedroom with the most direct/closest relationship to the potential through-floor lift route (see section 8.3) immediately adjacent to the accessible bathroom (as defined in section 7.6.2).
- In addition, where practicable, provide a full height 'knock-out' panel

in the wall between this bedroom and bathroom to enable the creation of a convenient direct connecting door between the two rooms if required in the future. The width of the 'knock-out' panel should enable provision of a minimum door opening width to accord with section 7.2.2.

1 **An accessible WC** (see Figure 14) with:
   a) a centre line that is between 400 mm and 500 mm from an adjacent wall.
   b) a flush control located between the centre-line of the WC and the side of cistern furthest away from the adjacent wall.
   c) an approach zone extending at least 350 mm from the WC's centre line towards the adjacent wall, and at least 1000 mm from the WC's centre line on the other side. This zone should extend forward from the front rim of the WC by at least 1100 mm. The zone should also extend back on one side of the WC for at least 500 mm from the front rim of the WC, for a width of 1000 mm, from the WC's centre line.

The bowl of a wash basin, which may be located either on the adjacent wall, or adjacent to the cistern, should not project into this approach zone by more than 200 mm.

2 **A wash basin** (see Figure 15) with:
   a) a clear frontal approach zone, 700 mm wide, extending 1100 mm from any obstruction under the basin's bowl, such as a pedestal, trap, duct or cabinet furniture. This zone will normally overlap with the approach zone to the WC (see **1c**), and/or bath approach zone (see **3**).

3 **A bath, or an accessible floor-level shower, or both:**
   a) where a bath is provided, there should be a clear zone alongside the bath that is at least 1100 mm long and 700 mm wide. This zone will normally overlap with the approach zone to the WC (see **1c**) and/or the approach zone to the basin (see **2**).
   b) where an accessible floor-level shower is provided instead of a bath, there should be a clear 1500 mm diameter circular, or 1700 mm × 1400 mm elliptical, clear manœuvring zone — see **5**). This manœuvring zone should overlap with the showering area. The drainage for the shower should be provided by shallow falls in the floor, to drainage. The fall gradients should be the minimum required to drain the catchment area of the shower effectively. Crossfalls should be minimised (see **4**).
   c) where both a bath and an accessible floor-level shower are provided from the outset, the clear floor space for showering activity should be a minimum of 1000 mm × 1000 mm. The drainage for the shower within this space should be as detailed in **4**.

4 Unless already provided (see **3b** and **3c**) or provided elsewhere in the dwelling (see section 6.4.5), **floor drainage** for a future accessible floor-level shower with:
   a) a floor construction that provides either shallow falls to the floor drainage, or (where the drainage is initially capped for use later following installation of a shower) that allows simple and easy provision of a laid-to-fall floor surface in the future. Where future provision would involve incorporation of an accessible shower tray within the floor construction (eg by breaking into a screed), the tray should have shallow falls. Its edges should be 'level' with the surrounding floor area, with neither a step nor a ramp. Where future provision would require pumped drainage, all necessary service connections for the pump should be provided, at easily accessible locations, from the outset.
   b) the drainage, when capped for later use following adaptation, maybe located under a bath.
   c) fall gradients in the floor that are the minimum required to drain the catchment area of the shower effectively. Whether provided from the outset or by subsequent adaptation, crossfalls should be minimised.

In dwellings with more than one storey, if drainage for an accessible shower is provided elsewhere (eg on the entrance level as required by section 6.4.5), the floor drainage need not be provided within the bathroom. In these cases, the floor drainage in the bathroom becomes a recommendation.

5 **A manœuvring circle or ellipse:**
   a) where an accessible floor-level shower is provided instead of a bath, there should be a clear 1500 mm diameter circular, or 1700 mm × 1400 mm elliptical, clear manœuvring zone. This manœuvring zone should overlap with the showering area.
   b) potential for a clear 1500 mm diameter circular or 1700 mm × 1400 mm elliptical clear manœuvring zone (on removal of the bath) where a bath is provided with capped drainage for a future accessible floor level shower beneath it.

In either **a** or **b**, the manœuvring circle or ellipse may pass under a wash basin bowl but should be clear of any pedestal, trap, duct or cabinet furniture.

6 **A structure above the WC and bath/ shower positions** that is capable of supporting, or capable of simple adaptation to enable adequate support for, ceiling hoists.

7 **A wall construction** (see Figure 17) within the bathroom (and within all other WC and bathroom facilities throughout the dwelling) that is capable of immediate firm fixing and support for adaptations such as grab rails within a height band of 300–1800 mm from the floor, as and when these are required to assist with independent use of facilities (with no strengthening or adaptation to the wall structure being necessary).

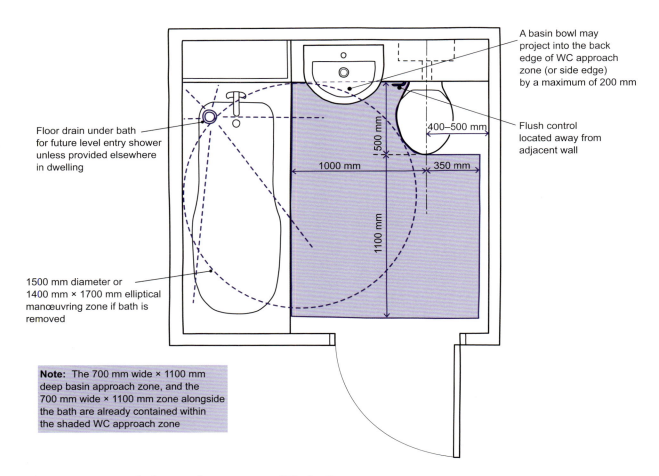

Figure 23: **Example layout for an accessible bathroom**

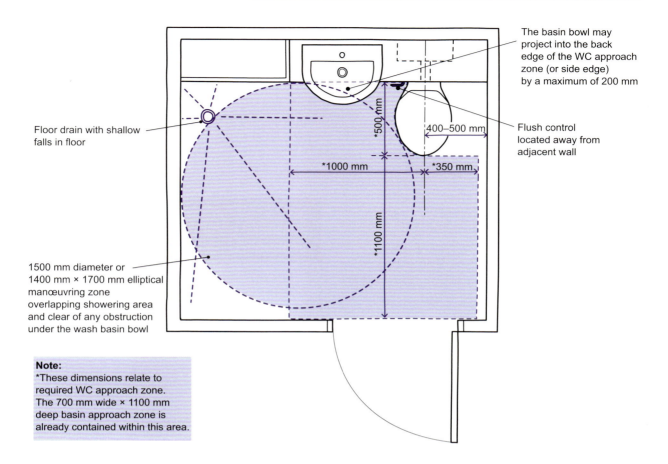

The basin bowl may project into the back edge of the WC approach zone (or side edge) by a maximum of 200 mm

Floor drain with shallow falls in floor

Flush control located away from adjacent wall

400–500 mm

*500 mm

*1000 mm

*350 mm

*1100 mm

1500 mm diameter or 1400 mm × 1700 mm elliptical manœuvring zone overlapping showering area and clear of any obstruction under the wash basin bowl

Note:
*These dimensions relate to required WC approach zone. The 700 mm wide × 1100 mm deep basin approach zone is already contained within this area.

Figure 24: **Example layout for an accessible shower room**

Figures 23 and 24 show example bathroom and shower room layouts that demonstrate the spatial requirements of the above items. It should be noted that an internal footprint dimension of 2100 mm × 2100 mm increases the degree of choice and flexibility in respect of fittings, layout, orientation and future adaptability. An outward opening door will be required to satisfy AD M[4] if the bathroom or shower room contains the only accessible entrance-level WC within the dwelling.

Additional good practice recommendations exceeding the minimum requirements for accessible bathrooms are given in Box 14.

Box 14: Accessible bathrooms/shower rooms — additional good practice recommendations

- Where possible, provide for a direct connection between the bathroom and a main bedroom. This will normally take the form of a full-height knock-out panel, capable of being fitted with a doorset, which achieves a clear opening in accordance with section 7.2.2.

- Provide floor drainage as described in item **4** above within the bathroom, even when it is provided elsewhere in the dwelling, to increase choice and convenience for adaptation and future use.

7.6.3 All bathrooms, en-suites and WCs

Clear door opening width

Every bathroom, en-suite and WC compartment within a dwelling should have minimum clear opening door widths as specified in section 7.2.2.

Wall construction

All walls (and boxing and ductwork) in every bathroom, en-suite and WC compartment within a dwelling should have a construction capable of immediate firm fixing and support for adaptations such as grab rails within a height band of 300–1800 mm from the floor. Immediate firm fixing of these items may be required at any point within this height band, and this should be possible without the need for any strengthening or wall adaptation. For example, timber frame stud walls should be lined with 18 mm plywood, or reinforced with 18 mm ply between all studs (see Figure 17).

An additional good practice recommendation exceeding the minimum requirements for all bathrooms and en-suites is given in Box 15.

Box 15: All bathrooms and en-suites — additional good practice recommendation

- Provide as many of the approach zones, and accessible features, described in sections 6.4 and 7.6.2, as practicable, in any additional bathrooms and WCs within a dwelling (in addition to the required accessible bathroom and entrance-level accessible WC).

8 CIRCULATION BETWEEN STOREYS WITHIN THE HOME

8.1 INTRODUCTION

The design of a Lifetime Home will consider how to enable effective and efficient adaptation works to allow access to storeys above the entrance level for household members with less agility and mobility. These adaptations relate to an ongoing or permanent need that cannot be catered for by the entrance-level facilities discussed in chapter 6.

A minimum stair width will help with assisted movement on the stairs, and ensure convenient space for a stair-lift, if and when a household member needs one.

A stair-lift will not always provide adequate means of access to the storey above. Early layout considerations should therefore also ensure that a convenient route is available to enable provision of a through-floor lift, by an efficient and cost-effective adaptation, so that convenient access to the accessible bathroom and a main bedroom is always possible.

Potential access to storeys above the entrance level of a dwelling, either via the stairs, a stair-lift, or a through-floor lift, should therefore be possible to cater for a wide range of existing or potential needs.

8.2 STAIRS

8.2.1 Application

Requirements 8.2.2 and 8.2.3 apply to all staircases within a dwelling.

8.2.2 Form of stairs

In dwellings with two or more storeys, all stairs and the associated area should enable a (seated) stair-lift to be fitted in the future without significant alteration or reinforcement.

8.2.3 Stair width

A clear width of 900 mm, measured 450 mm above the pitch line, should be provided on all staircases within a dwelling (Figure 25). This width will:
- assist with movement past a 'parked' stair-lift
- aid assisted movement on the stairs
- be more convenient for parents carrying children
- assist generally with movement of furniture and objects between storeys.

Additional good practice recommendations exceeding the minimum requirements for stairs are given in Box 16.

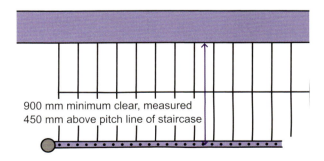

900 mm minimum clear, measured 450 mm above pitch line of staircase

Figure 25: **Stair width within dwellings**

Box 16: Stairs — additional good practice recommendations

- Provide a straight flight of stairs, without winders. Although stair-lifts are available for most forms of stair, a straight flight will provide for a more cost-effective stair-lift provision. A straight flight with goings of consistent depth with no winders is safer to use, particularly for those who are less agile.

- Provide a straight flight of stairs with a clear landing and wall of minimum 200 mm beyond the top nosing, and a clear landing and wall of minimum 700 mm beyond the bottom nosing to provide for a most cost-effective and simple stair-lift provision. Clear landings and walls at the top and bottom of the stairs are preferred, and recommended, although stair-lifts with hinged rails can provide a solution where these are not available.

8.3   ROUTE FOR A FUTURE THROUGH-FLOOR LIFT

8.3.1 Provision

A potential route for a through-floor lift should be identified in all dwellings unless the entrance level of the dwelling contains the living accommodation, the kitchen, a main (twin or double) bedroom and an accessible bathroom meeting the requirements of 7.6.2. Where required, this route should enable potential access, via the lift, to those rooms listed in the preceding sentence that are not on the dwelling's entrance level.

8.3.2 Potential route

The identified route for the lift may be from a living room/space directly into a bedroom above. Alternatively, the route may be from, or arrive in, a circulation space (Figure 26).

If the lift is to arrive into a bedroom, note the additional guidance in 8.3.3.

Potential route clear of services

To minimise adaptation costs and disruption, the potential lift route and the potential lift entrances and exits should be clear of services.

Acceptable degree of wall adaptation

It is acceptable for the identified route to require some degree of alteration or moving of lightweight partition walls (eg timber stud walls) if this can provide the most efficient and practical layout arrangement following lift installation. However, where this is the case, no services should be located within or adjacent to the wall(s) requiring alteration.

8.3.3 Requirements if the potential arrival point is directly into a bedroom

When the potential arrival point for the lift is in a bedroom, there must be space to exit and approach the lift. A compromised room layout would be expected following lift installation, but as a basic minimum the room should still be able to function as a single bedroom. However, if the lift route is to arrive directly into a bedroom, which then functions as a single bedroom, at least one other bedroom on that storey, should be a main (double or twin) bedroom (see section 8.3.1).

The most convenient arrangement is typically enabled when the lift arrives into the main bedroom satisfying sections 7.5.1 and 7.5.2.

Basic bedroom furniture provision for bedroom use must be possible in all instances.

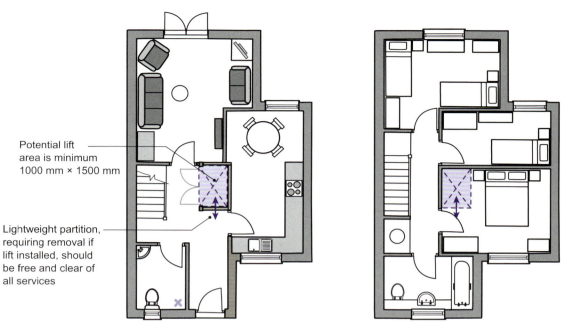

Potential lift
area is minimum
1000 mm × 1500 mm

Lightweight partition,
requiring removal if
lift installed, should
be free and clear of
all services

Potential through-floor lift route utilising storage space on ground floor

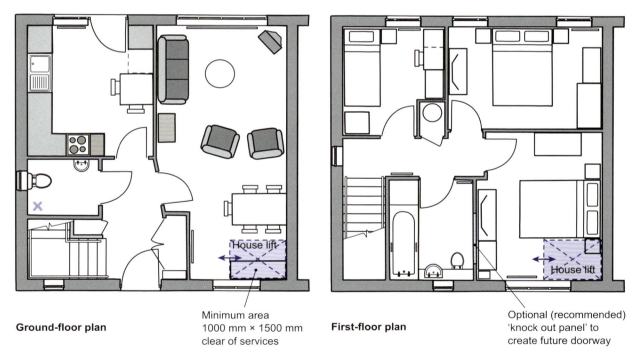

Ground-floor plan

Minimum area
1000 mm × 1500 mm
clear of services

First-floor plan

Optional (recommended)
'knock out panel' to
create future doorway

Potential through-floor lift route, living room–bedroom, requiring moderate adjustment of optimum furniture layout

Figure 26: **Examples of potential through-floor lift routes**

8.3.4 Aperture

Aperture size and orientation
The potential aperture size for the route through the floor should be a minimum of 1000 mm × 1500 mm, with the potential approach and exit to/from the lift being on one of the shorter sides.

Free of services
The potential aperture area should be clear of services.

Pre-formed 'knock-out' panel
If the identified lift route within the dwelling passes through a concrete floor, a 'knock-out' panel should be pre-formed within the floor. Traditional wooden joist floors, 'I' beam floors, and metal web floors need not be provided with a 'knock-out' panel along the lift route, provided that their design has taken account of

associated point loads to enable the creation of the void if required.

An additional good practice recommendation exceeding the minimum requirements for routes for a future through-floor lift is given in Box 17.

Box 17: Route for a future through-floor lift — additional good practice recommendation

- Provide an electrical point on the wall adjacent to the potential route to assist in any future installation of the lift. This plate should be annotated with 'lift position' (or similar) to assist in future identification of the possible route.

---

9 SERVICE AND VENTILATION CONTROLS

9.1 INTRODUCTION

The position of service controls that are used regularly, or may be needed in an emergency, should be positioned so that potential use is possible by a wide range of household, and potential household, members.

Similarly, to enable simple control of ventilation, one opening light to a window in each room should maximise potential access to the widest range of occupants.

9.2 SPACE TO APPROACH A WINDOW IN HABITABLE ROOMS

There should be potential for an approach route, 750 mm wide, to enable members of the household, including those that may be using a wheelchair, to approach an opening light within a window in each habitable room. This opening light should have handles no higher than 1200 mm from the floor. To provide the potential approach route to the window, some bedrooms may therefore require a minimum width (if it will be necessary to pass the foot of the bed in order to approach the window; see section 7.5.3).

9.3 WINDOW HANDLE HEIGHTS

9.3.1 Habitable rooms

In habitable rooms, the approachable window (see section 9.2), should have handles to an opening light no higher than 1200 mm from the floor.

9.3.2 Non-habitable rooms

Kitchen areas, bathrooms or other non-habitable rooms with windows positioned behind fittings and fixtures, have no requirement for a potential clear approach space to those windows. However, the maximum window handle height of 1200 mm for one opening light within the room remains applicable. Where a second window is located away from fixtures and fittings, this should, whenever possible, have the opening light with a maximum handle height of 1200 mm.

9.4 ELECTRICAL CONTROLS

Sockets, light switches and isolating switches to frequently used appliances (eg cooker switches), TV, telephone and computer points should be located within a height band of 450 mm–1200 mm from the floor and a minimum of 300 mm from any internal corner (Figure 27).

Switches within consumer service units (CSU) should also be within the height band of 450 mm–1200 mm from the floor (Figure 27).

9.5 CENTRAL HEATING CONTROLS

Programmer controls (whether located at the boiler or remote from the boiler), and any thermostatic or temperature controls should be located within a height band of 450 mm–1200 mm from the floor and a minimum of 300 mm from any internal corner (Figure 27). This would therefore include radiator temperature control valves, which may need to be sited towards the top of radiators.

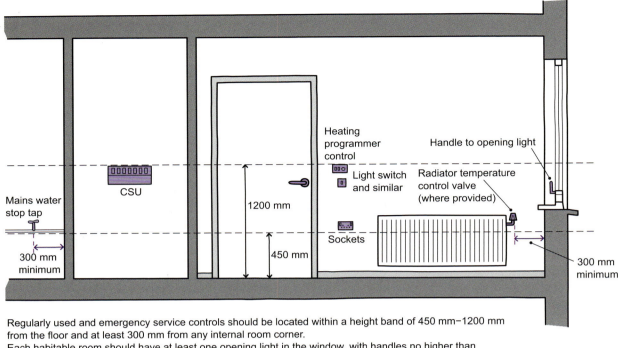

Regularly used and emergency service controls should be located within a height band of 450 mm–1200 mm from the floor and at least 300 mm from any internal room corner.
Each habitable room should have at least one opening light in the window, with handles no higher than 1200 mm from the floor.

CSU = Consumer service unit.

Figure 27: **Emergency and frequently used service control height band**

9.6 MAINS WATER STOP TAPS AND CONTROLS

Stop taps, or similar, used to shut down water flow in an emergency should be located within a height band of 450 mm– 1200 mm from the floor and a minimum of 300 mm from any internal corner (Figure 27).

Additional good practice recommendations exceeding the minimum requirements for service and ventilation controls are given in Box 18.

Box 18: **Service and ventilation controls — additional good practice recommendations**

- Locate different types of service control within the more specific height bands as detailed in Figure 26 of BS 8300:2009[9] to provide greater convenience for some household members.

- Wherever possible, locate similar controls in consistent locations throughout the dwelling, and provide them with a tonal contrast against their surroundings.

- Provide fused spurs to assist with potential future adaptations as detailed in previous chapters (eg future provision of stair-lift, through-floor lift, showers). This will enable more cost-effective adaptation, if and when these facilities become required.

- Provide taps with short lever operation, or of other designs, that will be operable by people with less finger and hand dexterity.

APPENDIX
CHECKLIST OF REQUIREMENTS

- Cross-reference should be made to the detailed specification within the main text of the design guide for detailed design requirements.
- This checklist relates to minimum requirements, and does not include any additional good practice recommendations.

3   APPROACHING THE HOME

3.2   PARKING

3.2.2 Parking dimensions for communal parking
- Where communal parking is provided is one space with the required minimum dimensions located close to each block entrance or core?

3.2.3 Parking width for 'private' (on plot) parking
- Where private 'on-plot' is provided for an individual dwelling, does one space have the required minimum width, or potential to have it?

3.2.4 Gradients and surfaces of parking spaces
- Are the spaces (referred to in 3.2.2 and 3.2.3) 'level'?
- Do the spaces (referred to in 3.2.2 and 3.2.3) have suitable surfaces?

3.2.5 Car ports
- If a car port provides the only parking space for a dwelling, does it meet the minimum width, gradient and surface requirements?

3.3   ACCESSIBLE ROUTE(S) BETWEEN PARKING AND DWELLINGS OR BLOCKS OF DWELLINGS

3.3.1 Provision of accessible route(s)
- Is the principal access route to the relevant entrances from these parking spaces an accessible route (see 3.3.3–3.3.6 below)?

3.3.2 Relevant entrances for accessible route(s)
- Where the principal access route is to a block of dwellings, is it to the block's main entrance, or (in the case of basement parking) to the lift core entrance?
- Where the principal access route is to an individual dwelling, is it to both the dwelling's principal entrance and any secondary entrance? Where it is to only one entrance of the dwelling can it be demonstrated that topography and/or regulation prevent an accessible route to both entrances?
- Where the accessible route is to only one entrance of the dwelling, is it to the principal entrance? If it is to a secondary entrance, can it be demonstrated that topography and/or regulation prevent the accessible route being to the principal entrance?

- If the accessible route is to only one entrance of the dwelling, is this the entrance closest to the parking spaces discussed in 3.2.3 above?

3.3.3 Distance of accessible route(s)
- Are the accessible routes as short as practicable?

3.3.4 Gradients along the accessible route(s)
- Do the accessible routes meet the required maximum gradient/distance ratios?
- Do any slopes have top and bottom level landings, and any required level resting areas?

3.3.5 Width of accessible route(s)
- Do these accessible routes have the required minimum width?

3.3.6 Level landings at external entrance(s)
- Do the accessible routes have a level landing of the required size adjacent to the associated entrance doors?

3.3.7 Surface of accessible route(s)
- Do the accessible routes have suitable surfaces?

3.3.8 Additional stepped approach
- If a principal accessible route to a communal entrance involves slopes/ramps is there also an additional stepped approach in accordance with AD M[4]?

3.4 OTHER APPROACH ROUTES TO DWELLINGS (FROM OTHER PARKING OR FROM THE HIGHWAY)

3.4.1 Gradients on other approach routes
- Do all other approaches to entrances from any parking, or highway, have suitable gradient/distance ratios? If not, is the site steeply sloping? If the site is steeply sloping, have the proposed details for these other approaches (which are in addition to the required accessible routes discussed in 3.3.1–3.3.6) been discussed, and agreed, with the local planning authority?

3.4.2 Width on all approach paths
- Do these other approach paths also have the required minimum widths?

3.4.3 Surface treatments on all approach paths
- Do these other approach paths also have suitable surfaces?

---

4 ENTRANCES

4.2 EXTERNAL LIGHTING

4.2.1 Application
- Does every external entrance have fully diffused external lighting?

4.3 ACCESSIBLE THRESHOLDS

4.3.1 Application
- With the exception of 'Juliet' balconies and balconies/roof terraces over habitable rooms, do all external entrances (including all other balconies and roof terraces) have accessible thresholds?

4.3.2 Maximum up-stand
- Do the accessible thresholds have a maximum total up-stand of 15 mm (consisting of a number of smaller up-stands and sloping infill connection between)?
- Is the slope on any sill between the threshold and the external surface 15° or less?

4.3.3 Internal transition units
- If an internal transition unit is provided does it have a slope of 15° or less?

4.4 EFFECTIVE CLEAR OPENING WIDTHS OF ENTRANCE DOORS

4.4.1 Dwelling entrance doors
- Do all dwelling entrance doors have the required effective minimum clear opening widths?

4.4.2 Communal entrance doors
- Do all communal entrance doors have the required effective minimum clear opening widths (according to the width and direction of approach)?

4.5 NIBS

4.5.1 Application
- Do all entrance doors have the required nib?

4.6 IRONMONGERY AND ACCESS CONTROLS

4.6.1 Height
- Are all handles, locks and other access controls within the required height band?

4.6.2 Location
- Are all handles, locks and other access controls located away from the corner of any side wall return?

4.7 EXTERNAL LEVEL LANDINGS AT MAIN ENTRANCES

4.7.1 Application
- Are level landings provided at all relevant entrances?

4.7.2 Dimensions
- Do the level landings have the required dimensions?

4.8 WEATHER PROTECTION AT MAIN ENTRANCES

4.8.1 Covers and canopies
- Is overhead weather protection provided at relevant entrances?

4.8.2 Size and form of cover
- Is the size of the cover adequate for local conditions, and the position of entry controls?

5 INTERNAL CIRCULATION WITHIN COMMUNAL AREAS

5.2 INTERNAL COMMUNAL DOOR WIDTHS
- Do all communal doors have the required minimum clear opening widths, depending on the direction of approach to each door and the width of that approach?

5.3 NIBS

5.3.1 Provision
- Do all communal doors have the required nib to the leading edge on the pull side?

5.4 COMMUNAL HALLWAY, CORRIDOR AND LANDING WIDTHS
- Do all communal hallways, corridors and landings have the required minimum widths in accordance with the clear opening widths provided to doorways in their side walls?
- If there are no doorways in the side walls of a communal hallway, corridor or landing, does the hallway, corridor or landing have the required minimum width?
- If there is a 'pinch point' in a communal hallway, corridor or landing, is the required minimum width available at the 'pinch point', and is the obstacle causing the 'pinch point' away from doorways and any change of direction?

5.5   COMMUNAL STAIRS

5.5.1 Pitch/5.5.3 Application
- Do all communal stairs that provide a principal access route to a dwelling, regardless of whether or not a lift is provided, have an 'easy going' pitch?

5.5.2 Handrails
- Is the height and extension of handrails on communal stairs as required?

5.6   COMMUNAL PASSENGER LIFTS

5.6.2 Size of lift
- Where provided, do passenger lifts have the required minimum internal dimensions?

5.6.3 Position of lift controls
- Where provided, do passenger lifts have correctly positioned controls?

5.6.4 Landings
- Are adequately dimensioned landings provided adjacent to passenger lifts on all storeys served by the lift(s)?

---

6   ENTRANCE-LEVEL FACILITIES WITHIN THE HOME

6.1   INTRODUCTION
- Does the perceived 'entrance level' of the dwelling accord with the definition of 'entrance level' for Lifetime Homes purposes?

6.2   LIVING SPACE

6.2.1 Provision
- Is there a permanent living room/space on the entrance level?

6.2.2 Sight line through windows of principal living spaces
- If the living space in 6.2.1 is the principal living space, does its main window/glazing area achieve the required glazing line height?
- Does this window also offer a suitable sight line to the outside from a seated position?
- If this living space is not the principal living space, is the required glazing line height, and sight line from a seated position, provided in the principal living space?

6.3   POTENTIAL FOR AN ENTRANCE-LEVEL BED-SPACE

6.3.1 Provision/6.3.2 Size of space
- Unless there is a bedroom on the entrance level, is there a suitable area on the entrance level that could be used as a temporary bed-space?

6.3.2 Provision of electrical socket
- Does this potential temporary bed-space have an electrical socket?

6.4   PROVISION OF AN ACCESSIBLE WC, BASIN AND SHOWER DRAINAGE

6.4.1 Provision of floor drainage for and a future accessible shower
6.4.5
- Do all dwellings have floor drainage on their entrance level to enable a future accessible shower?

6.4.3 Accessible WC
6.4.4 Accessible basin
- Do all dwellings, other than one or two bed houses/maisonettes (as defined in chapter 6), have an entrance-level accessible WC with:
  - an accessible WC at an acceptable distance from a side wall, a correctly positioned flush control and the required approach zone?
  - an accessible basin with the required approach zone?
  - appropriately located floor drainage and suitable floor construction?

6.4.6 Wall construction

Does the wall construction of the room containing the entrance-level WC enable immediate firm fixing of support rails within the required height band?

7   CIRCULATION AND ACCESSIBILITY WITHIN THE HOME

7.2   INTERNAL HALLWAYS, LANDINGS AND DOORWAYS

7.2.1   Minimum hallway and landing widths

- Do all hallways and landings within a dwelling have the necessary minimum widths (depending on the width and locations of associated door openings)?

7.2.2   Minimum internal doorway widths

- Do all the internal doorways that a person may pass through within the dwelling, have the required minimum clear opening width (depending on the approach direction and the width of that approach)?

7.2.3   Nibs

- Do all the internal doorways to rooms on the entrance level of each dwelling have the required nib to the leading edge/pull side?

7.3   LIVING ROOMS/AREAS AND DINING ROOMS/AREAS

7.3.1   Turning circle or ellipse

- Do living rooms/areas and dining rooms/areas have the required turning circles or ellipses?

7.3.2   Circulation space between items of furniture

- Do living rooms/areas and dining rooms/areas have minimum space between furniture on essential circulation routes?

7.4   KITCHENS

7.4.1   Space in front of kitchen units and appliances

- Is there the required clear distance in front of all kitchen units and appliances in the kitchen?

7.5   BEDROOMS

7.5.1   Circulation space within a main bedroom

- Does one main bedroom have the required clear width around both sides, foot and corners of a double bed?

7.5.2   Relationship of a main bedroom to the accessible bathroom/ potential for through-floor lift access and ceiling hoist

- Is a main bedroom close to an accessible bathroom (see 7.6.2) and on the storey with potential access from a through-floor lift (unless entrance level)?
- Is the structure above the ceiling over the main bedroom close to the accessible bathroom, and the ceiling over the accessible bathroom, capable of supporting (or capable of adaptation to support) ceiling hoists?

7.5.3   Circulation space within other bedrooms

- Can all other bedrooms have the required clear space to one side of the bed?
- Where it is necessary to pass the foot of the bed to approach the window, is the required clear width (to pass the foot of the bed) available?

7.6   BATHROOMS AND WCs

7.6.2 Accessible bathroom

- Does the dwelling have an accessible bathroom, close to, and on the same storey as a main bedroom, either on the entrance level or on the storey with potential access from a through-floor lift?
- Does this accessible bathroom have the following facilities:
  - an accessible WC at an acceptable distance from a side wall, a correctly position flush control and the required approach zone?
  - an accessible basin with the required approach zone?
  - either a bath, or an accessible floor shower (or both), with the required associated approach zone(s) and manœuvring zone(s)? Where a shower is provided, is the floor drainage and showering area as specified?
- Unless provided elsewhere in the dwelling, does this bathroom (even where no shower is provided from the outset) have drainage (possibly under a bath) to enable a level-entry accessible shower? Are there shallow falls in the floor to the drainage, or is there a floor construction that will enable easy provision of a laid-to-fall floor surface (and connection to existing drainage) in the future?

7.6.3 All bathrooms, en-suites and WCs

- Does every bathroom, en-suite and WC compartment have the required minimum clear door opening widths (see 7.2.2).
- Does every bathroom, en-suite and WC compartment have a wall construction capable of immediate firm fixing and support for grab rails and similar at any location within the required height band?

8   CIRCULATION BETWEEN STOREYS WITHIN THE HOME

8.2   STAIRS

8.2.2 Form of stairs

- Do all staircases within the dwelling enable future fitting of stair-lifts without the need for significant alteration or reinforcement?

8.2.3 Stair width

- Do all staircases have the required minimum width?

8.3   ROUTE FOR A FUTURE THROUGH-FLOOR LIFT

8.3.1 Provision

- In dwellings of more than one storey, is a potential route for a through-floor lift identified that connects storeys containing the necessary rooms?

8.3.2 Potential route

- Are the potential lift route and potential lift entrance and exit routes clear of all services?
- If the potential lift route requires moderate alteration of lightweight partitions, are these partitions clear of all services?

8.3.3 Requirements if the potential arrival point is directly into a bedroom

- If the lift is to arrive into a double bedroom can this still function as a double room (with a compromised room layout) or is there another main (double or twin) bedroom on the same storey?
- If the lift arrives into a bedroom which then needs to function as a single bedroom, is there at least one other bedroom on that storey which can function as a main (double or twin) room?

8.3.4 Aperture

- Is the identified potential lift aperture area of the minimum required size and of a suitable orientation to enable access?
- Is the identified aperture area clear of services?
- If the potential lift route passes through a concrete floor, is a 'knock-out' panel to be pre-formed within the floor structure?
- Where necessary, has the floor design taken account of the potential loadings associated with the possible lift void?

9   SERVICE AND VENTILATION CONTROLS

9.2   SPACE TO APPROACH A WINDOW

With the exception of kitchens and bathrooms that have all windows behind fixtures and fittings, is there potential for an approach route of the minimum required width to an opening light of a window in each room? Are handles on this opening light no higher than 1200 mm from the floor?

9.3   WINDOW HANDLE HEIGHTS

- Do all rooms with a window have at least one opening light with handles no higher than 1200 mm from the floor?

9.4   ELECTRICAL CONTROLS

- Are all frequently used electrical controls within the required height band and the minimum distance away from any internal corner?

9.5   CENTRAL HEATING CONTROLS

- Are central heating programmer controls and any thermostatic or temperature controls within the required height band and the minimum distance away from any internal corner?

9.6   MAINS WATER STOP TAPS AND CONTROLS

- Are controls such as main stop taps (used to shut down water flow in an emergency) located within the required height band and the minimum distance away from any internal corner?

REFERENCES

1 Lifetime Homes. Lifetime Homes Standard: 16 design criteria. Revised 2010. www.lifetimehomes.org.uk.

2 Thorpe S and Habinteg Housing Association. Wheelchair housing design guide. Bracknell, IHS BRE Press, 2006, 2nd edn.

3 Greater London Authority (GLA). Wheelchair accessible housing: designing homes that can be easily adapted for residents who are wheelchair users. London, GLA, 2007.

4 Department for Communities and Local Government (CLG). The Building Regulations (England & Wales) 2010. Approved Document M: Access to and use of buildings (2006 edn). London, CLG, 2006.

5 Greater London Authority (GLA). Accessible London: Achieving an inclusive environment. The London Plan Supplementary Planning Guidance. 2004

6 Department for Communities and Local Government (CLG). Lifetime Homes, Lifetime Neighbourhoods: A national strategy for housing in an ageing society. London, CLG, 2008

7 BSI. Design of accessible housing. Lifetime home. Code of practice. DD 266:2007. London, BSI, 2007.

8 DETR. Accessible thresholds in new housing – Guidance for house builders and designers. London, TSO, 1999.

9 BSI. Design of buildings and their approaches to meet the needs of disabled people. Code of practice. BS 8300:2009+A1:2010. London, BSI, 2009, 2010.

INDEX

OTHER TITLES FROM IHS BRE PRESS

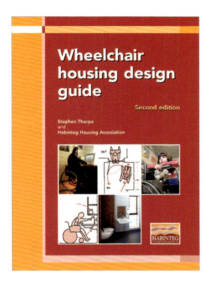

Learn how to design and detail a home that is fully manageable by wheelchair users and maximises their independence. This activity-based guide:

- discusses design considerations, requirements and recommendations
- provides design solutions
- promotes strategies for social inclusion and cost benefits of designing to wheelchair accessibility standards.

Over 150 detailed diagrams.

Ref. EP 70

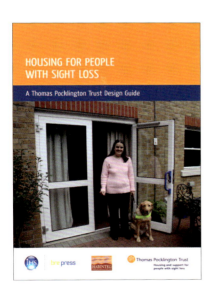

Get good practice guidance on the development of inclusive domestic environments to meet the requirements of people with sight loss. The emphasis is on maximising functional vision and minimising barriers and risk by achieving specific design and specification requirements.

Ref. EP 84

Order now @ **www.brebookshop.com** or **phone the IHS Sales Team on +44 (0) 1344 328038**.

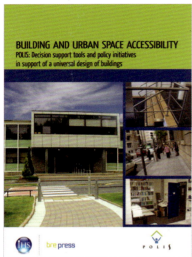

Learn about POLIS, a quantitative methodology developed for access auditing of buildings and urban spaces. This report:
- sets out the key features of the POLIS approach to the universal design of buildings
- is based on expert input from European experts
- includes five case studies from Spain, UK, Italy, Greece and Israel illustrating the POLIS assessment tool.

Ref. EP 83

Understand the basic requirements for accessible buildings which in making buildings more functional will be to the benefit of all users. This BRE Digest:
- outlines the requirements of national building regulations and the Disability Discrimination Act
- summarises access requirements inside and outside domestic and non-domestic buildings
- discusses communication devices, access auditing and assistive technology.

Ref. DG 505

Order now @ **www.brebookshop.com** or **phone the IHS Sales Team on +44 (0) 1344 328038**.

Coventry University Library